OUR VIRAL FUTURES

SCIENCE NOW

SCIENCE NOW

VINCENT ANTONIN LÉPINAY AND MATTHEW L. JONES,
SERIES EDITORS

Scholars who study science explore the plurality and contingency of scientific and technical knowledge and practice with evidence from social, economic, technological, and political contexts. Empirical analysis of scientific and technical practice, coupled with theoretical depth, characterizes the best work in the field, whether from history, anthropology, sociology, philosophy, or communication. Understanding how scientific knowledge comes about reveals the strengths and weaknesses of that knowledge, as well as its future value and veracity—the polar opposite of antiscience.

Publishing cutting-edge work in the field, Science Now is a methodologically diverse, editorially driven series focused on scientific, technological, and medical developments across the globe in the last 150 years. The series seeks books that reorient current conversations through the introduction of theoretically generative, empirically driven evidence and cases. The series also invites books by scholars who want to write about science with the rigor of historians while engaging with sophisticated philosophical and methodological questions. These works will be of interest to scholars across history, sociology, communication, philosophy, and anthropology—the wide-ranging fields that compose science studies. Scholars invited to contribute manuscripts to Science Now will be expected to critically analyze the novel political lives and structures that science and technology have ushered in.

OUR VIRAL FUTURES

A POLITICAL ECOLOGY OF MICROBES

CHARLOTTE BRIVES

Foreword by

BRUNO LATOUR

Columbia University Press   *New York*

Columbia University Press
Publishers Since 1893
New York   Chichester, West Sussex
cup.columbia.edu

Originally published as *Face à l'antibiorésistance. Une écologie politique des microbes* © Éditions Amsterdam, 2022
Translation © 2026 Charlotte Brives
Translated by Dean Frances for Hancock Hutton Language Services
All rights reserved

Library of Congress Cataloging-in-Publication Data
Names: Brives, Charlotte, 1981–, author | Latour, Bruno writer of foreword | Frances, Dean, translator
Title: Our viral futures : a political ecology of microbes / Charlotte Brives ; foreword by Bruno Latour.
Other titles: Face à l'antibiorésistance. English
Description: New York : Columbia University Press, [2026] | Series: Science now | "Originally published as Face à l'antibiorésistance. Une écologie politique des microbes © Éditions Amsterdam, 2022. Translated by Dean Frances for Hancock Hutton Language Services" | Includes bibliographical references and index.
Identifiers: LCCN 2025026172 | ISBN 9780231221979 hardback | ISBN 9780231221986 trade paperback | ISBN 9780231564007 epub | ISBN 9780231564953 PDF
Subjects: LCSH: Bacteriophages | Microbial ecology | Drug resistance in microorganisms | Evidence-based medicine
Classification: LCC QR342 .B75 2026
LC record available at https://lccn.loc.gov/2025026172

Cover design: Elliott S. Cairns
Cover illustration: Victoria Denys

GPSR Authorized Representative: Easy Access System Europe, Mustamäe tee 50, 10621 Tallinn, Estonia, gpsr.requests@easproject.com

To the list of agents he had added an element, the microbe, that was to play a crucial role in the rearrangement of all modes of life. Pasteur's case proves once again that science proceeds not through the simple expansion *of an already existing "scientific worldview" but through the* revision *of the list of objects that populate the world, something that philosophers normally and rightly call a* metaphysics *and that the anthropologists call a* cosmology.

Bruno Latour, *Facing Gaia*

Struggling against Gaia makes no sense—it is a matter of learning to compose with her. Composing with capitalism makes no sense—it is a matter of struggling against its stranglehold.

Isabelle Stengers, *In Catastrophic Times*

CONTENTS

FOREWORD

BRUNO LATOUR

Since the COVID-19 pandemic, many of us have been wondering how to learn from this event. It would be a real shame to waste the opportunity. Once we have gotten over the strange idea of a "war" against viruses, we are going to have to get used to "living with them." This is what Charlotte Brives seeks to explore by drawing her readers into her investigation.

However, among the vast number of viruses, it is necessary to choose those that lend themselves to a detailed analysis of interest to all those who have had to undergo, directly or indirectly, the ordeal of the epidemic. By chance, or rather by that singular tact that marks a first-rate researcher, the author has stumbled upon one of the most studied types of viruses, yet one whose status is most uncertain and, strangely enough, whose politics are, for the moment, wide open.

Learning to live with this particular being means delving into the mysteries of biology, questioning its ontological status and grasping a crucial moment in its collective existence. There's an opportunity here; whole swathes of medicine depend on it; nobody yet knows how the story will unfold. Well, that's just the place for a researcher bold enough to take part in the movement she's about to describe. Over the course of her book, Charlotte

Brives accompanies us through this whirlwind of uncertainties, in the midst of which a collective seeks to come to terms with a being with an uncertain destiny.

This type of virus is a bacterial killer known as a *bacteriophage*.

As a great lover of science, I have always delighted in the article that announces the appearance of this bacterial killer to the learned world. Its birth announcement was a two-page contribution by Félix d'Hérelle in the *Comptes rendus de l'Académie des sciences*, at the height of the First World War, on September 10, 1917. Two pages long!

Happy were the days when one could introduce a major discovery without jargon, without innumerable acronyms, without dozens of pages of "materials and methods," and without "work packages" to be distributed in advance in interminable grant applications. All of a sudden, D'Hérelle goes for a breakthrough: "On an invisible microbe antagonistic to dysenteric bacilli." He doesn't see it—which is why it is called a virus, that is, something that no filter can pick up because it is so tenuous—but he shows, in a few experiments typical of the great Pasteurian gesture, what this little bug does to the immense bacteria when cultivated with them. "Antagonist" is a euphemism: The invisible microbe makes them explode; it lyses them, as microbiologists say (my French corrector suggests "lèse," which isn't bad either).

D'Hérelle is not a strict Pasteurian. To say the least. His career is a novel. But on the essential point—the kind of existence led by the being he has just brought to light—he is faithful to his masters. It is a common misconception that Pasteur was a killer, an eradicator of microbes. On the contrary, he was always sensitive to the subtle interplay of various agents that interfere with one another, sometimes attenuating, sometimes increasing their virulence.

D'Hérelle immediately understood how delicate it was going to be to qualify the modus operandi of this microbe he triumphantly named "bacteriophage" at the end of his article. And he wrote this sentence, which the author could have put at the head of her book: "The antagonistic action is therefore not inherent in the very nature of the invisible microbe, but acquired in the patient's organism by cultivation in symbiosis with the pathogenic bacillus." Every word counts. Incredible prescience: The bacillus and the virus that destroys it live in symbiosis and learn from each other! That's enough to put clinicians' brains to work for a century. . . .

And it is precisely this reciprocal uncertainty about the status of these commensals that Charlotte Brives revisits a hundred years later. Two major events have occurred in the life of these bacteriophages since they first entered the lives of medical communities. On one hand, they have become simple tools for the nascent molecular biology laboratories. Under the abbreviated name of "phages," they serve as carriers of biochemical products and almost as syringes for researchers exploring DNA. As simple tools, their status no longer interested many people. More importantly, antibiotics have become so effective at chemically killing bacteria that the whole subtle dance between phages and "their" bacteria—or bacteria and "their" phages—no longer interests many people.

Until, with a reversal so typical of the effects of unbridled modernization, the success of antibiotics created the very conditions for their ineffectiveness! Antibiotic resistance is developing to such an extent that bacteriophages, having emerged from their silence and released from their role as mere DNA syringes, are once again becoming central, or could become so, or risk not becoming so. This is precisely what nobody knows. But this situation, seized on the fly, in the middle of the battle, is what Charlotte Brives puts before the reader's eyes. Suspense, suspense.

The role of a foreword is not to be a spoiler.

OUR VIRAL
FUTURES

INTRODUCTION

Antimicrobial Resistance: Global Report on Surveillance, a report published by the World Health Organization in 2014, makes for scary reading. In 256 apocalypse-tinged pages with supporting diagrams and figures, it reveals how antibiotics and other antimicrobials are gradually losing their effectiveness. According to Dr. Keiji Fukuda, assistant director-general for health security, who wrote the foreword, the problem is "so serious that it threatens the achievements of modern medicine. A post-antibiotic era—in which common infections and minor injuries can kill— . . . is . . . a very real possibility for the 21st century."[1] Pathogenic bacteria that once could have been treated with ease are now resisting human attempts to eradicate them from the bodies of infected people. Infections previously made benign can now strike people down again. Routine surgery, organ and tissue transplants, and chemotherapy will become riskier for patients as existing treatments become ineffective against the opportunistic infections common in these cases.

Some reports suggest that as many as ten million deaths per year worldwide will be caused by bacterial infections that have become resistant to antibiotics by 2050. However, there is no need to seek out models whose precepts can always be contested. Every day, new bacterial strains are isolated, containing ever-increasing numbers of genes that make them resistant to one, several, or—increasingly—all antibiotics currently available. The website for the French national health insurance fund, l'Assurance Maladie, currently states that 12,500 deaths per year in France are directly attributable to bacterial infections that antibiotics have failed to treat.[2] In a review published in January 2022, a consortium of scientists documented a clear picture of the situation in 2019 at the global level: More than one million deaths were directly caused by antibiotic-resistant bacteria. The authors also pointed out great disparities across geographic areas, with the African continent being hardest hit in terms of number of deaths per capita.[3]

In this critical context, calls for massive investment by pharmaceutical companies in the search for new antibiotics are multiplying, as are op-eds and opinion pieces calling for the creation of alternative development models.[4] Alternatives to antibiotics are also being developed and becoming increasingly widespread. Examples include the possibility of using antimicrobial peptides—small molecules found in the innate immune defense systems of most living things—or phage lysins—enzymes produced by bacteriophage viruses that have the ability to break down bacterial membranes.[5] However, another possibility now stands out in the biomedical landscape: using not molecules derived from biological entities but rather the biological entities themselves. This is the principle of phage therapy, that is, the use of bacteriophages—literally, "bacteria eaters"—to treat antibiotic-resistant infections. This book is devoted to this alternative solution, its implications, and the opportunities that it enables us to consider.

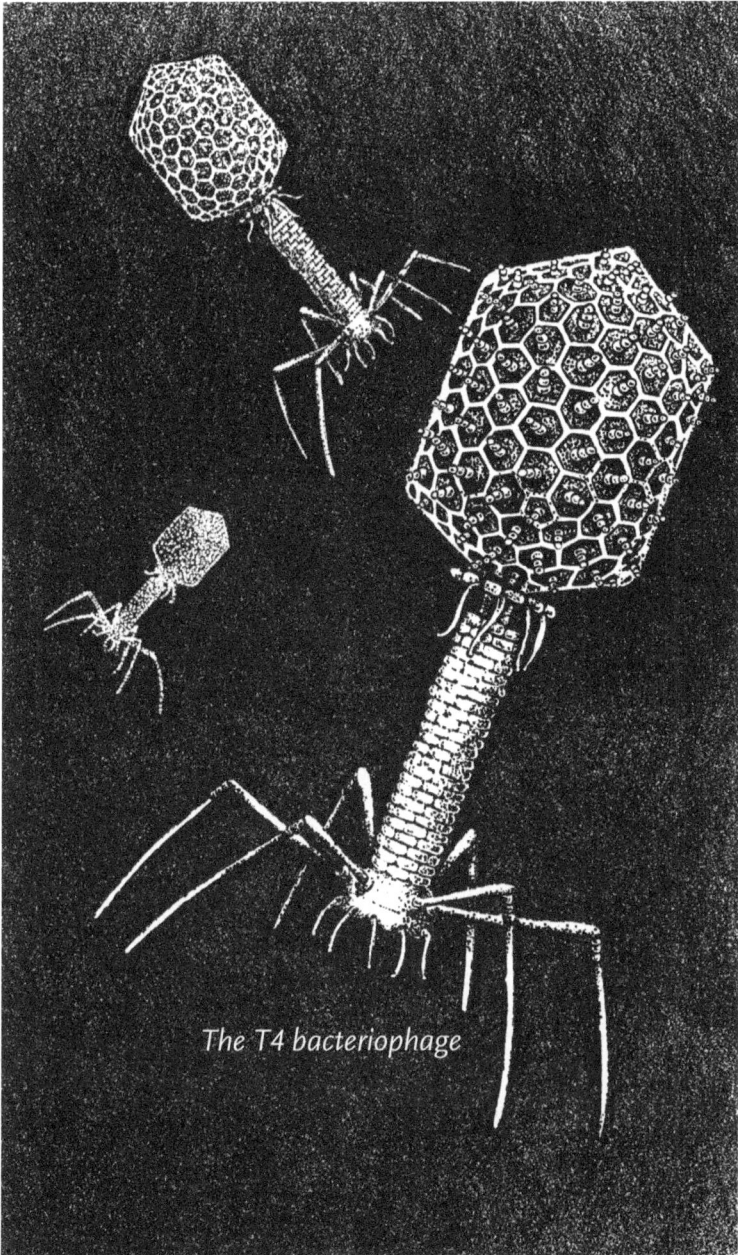

FIGURE 0.1 The T4 bacteriophage

THE PRINCIPLE OF
AN AGE-OLD THERAPY

The use of phages for healing purposes is an age-old practice that almost completely died out in Western Europe and North America over the second half of the twentieth century, partly because of the development of antibiotics, although it has persisted in certain forms in some countries of the former Eastern bloc, mainly Georgia, Russia, and Poland (chapter 2). The renewed interest in this therapy over the past fifteen years or so in France, as well as in Belgium and Switzerland, is clearly linked to the recognition of antimicrobial resistance (AMR) as a major public health problem. Interest is also being embodied locally in response to the particular and localized forms that resistance phenomena can take in patients' bodies, with each infection being a unique event (chapters 1 and 5). This practice, which remains extremely marginal in France despite its advocates' contagious enthusiasm, is still confined to limited contexts, mainly in the infectious disease departments of certain hospitals.

Bacteriophages, or phages, are viruses composed mainly of nucleic acids (DNA or RNA) and proteins that form an enveloping capsid (a type of shell). They are hosted by bacteria, which means that wherever bacteria are present, phages can also be found in abundance. For each bacterium on Earth, there are estimated to be around ten phages. For this reason, they are considered to be the most commonly occurring biological entity on the planet, with approximately 10^{31} (i.e., a one followed by thirty-one zeros) representatives. There are up to fifty million phages in a milliliter of seawater (a single drop of water), one hundred million in a gram of soil, and several million in a gram of stool.

Although bacteriophage viruses are usually described in publications intended for both a specialized readership and the general public as "professional killers" of bacteria, the relationship between the two is actually much more complex and has developed over the course of several billion years.[6] Phages and bacteria are entities in constant coevolution via various biological cycles. The most frequently observed, and therefore those about which scientists have gained the most knowledge, are the lysogenic and lytic cycles (figure 0.2). These two cycles, mentioned throughout this book, provide the keys to understanding the relationships between phages and bacteria and enable us to comprehend and analyze, as precisely as possible, the implications of using these viruses in medicine.

Encounters between phages and bacteria always begin in the same way: After an initial *adsorption* phase, during which a phage "recognizes" a pattern on the membrane of a bacterium to which it can attach itself, the virus injects its DNA into the bacterial cell. This first step is extremely specific: A phage can normally recognize only one given bacterial species and sometimes only some of the genetic variants (strains) of that bacterial species. Two possibilities then arise according to the phage's capabilities. If the phage is *virulent*, it will use the bacterium's cellular machinery to replicate its DNA and manufacture its capsid proteins. In this case, several dozen or even several hundred new phages, or virions, will be produced and released into the environment after *lysis* (disintegration) of the bacterial membrane. These phages can then bind to other bacteria, giving rise to the principle of self-sustained lysis: As long as bacteria remain present, phages will continue to bind to them and multiply. This is called the lytic cycle. However, if the phage is *temperate*, its DNA can be integrated into the bacterium's own DNA, making this bacterium a carrier of new genes that may give it new capabilities that can be passed on to its descendants. This is the lysogenic cycle. However, as driven by certain environmental factors,

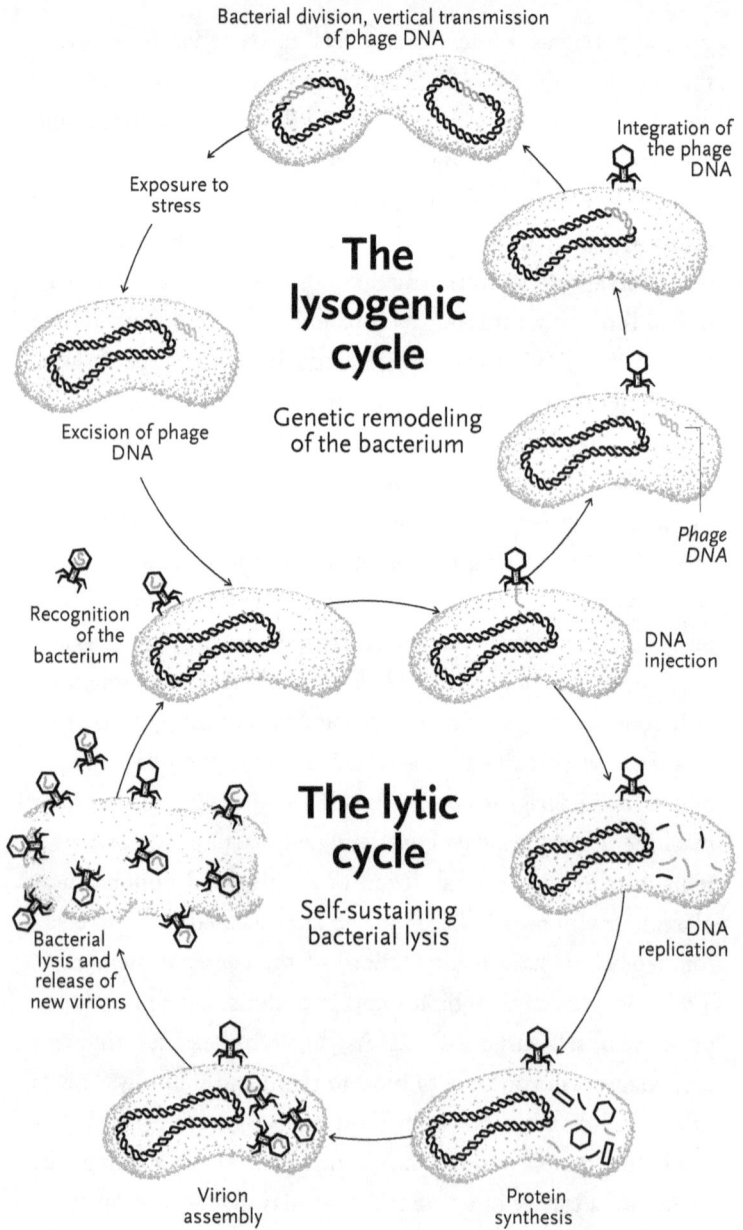

The lysogenic cycle

Genetic remodeling of the bacterium

Bacterial division, vertical transmission of phage DNA

Exposure to stress

Excision of phage DNA

Recognition of the bacterium

Integration of the phage DNA

Phage DNA

DNA injection

The lytic cycle

Self-sustaining bacterial lysis

Bacterial lysis and release of new virions

Virion assembly

Protein synthesis

DNA replication

FIGURE 0.2 The lysogenic and lytic cycles

the phage may excise itself: The phage DNA is detached from the bacterium's DNA, and a lytic cycle can then begin.

Phage therapy is based on using only certain potentialities of these viruses, namely their lytic properties, that is, their ability to destroy bacteria.[7] The process consists of selecting one or more virulent phages that are active on the bacteria responsible for the infection to be treated and administering them as close as possible to the location of the bacteria, thus enabling the adsorption phase—and therefore the process of self-sustained lysis—to take place. This simple and elegant principle underpins the use by humans of one of the many potentialities of microorganisms.

However, the complex history of phage therapy and its developmental difficulties show that there is nothing simple about its implementation. How can we establish the existence of these viruses as therapeutic entities? This is the question that the people mentioned in this book are trying to answer—while also addressing the problems it raises and the many challenges it poses.

This question should not be dissociated from the accumulated data on phage–bacteria relationships as summarized by the lysogenic and lytic cycles. The challenge for this book is to consider phage therapy in relation to the available knowledge and expertise while associating the materialities of societies with the materialities and agencies of living beings—human and nonhuman. Phages are *obligate* parasites, that is, biological entities that cannot reproduce without their bacterial hosts. They are deeply *relational* entities. However, phages and bacteria are also fluid entities, capable of merging with and transforming each other whenever they meet. As such, they can be defined only with difficulty and on an ad hoc basis according to the categories used by humans to describe living entities.

In this book, I set out to not only reveal the diversity and plasticity of the relationships among phages, bacteria, and humans but also describe how these relationships lead to their transformation by presenting an ethnography of the practices in which these entities are involved. In so doing, I explore how medicine, care, infection, and healing can, in turn, be transformed when these processes are no longer related to organisms considered in isolation but to networks of relationships.

The two cycles—lysogenic and lytic—can also be used to study a series of phenomena that require quite different scales of analysis and understanding. Although scientists are still a long way from possessing detailed knowledge of the roles played by the interactions between phages and bacteria, these interactions are nevertheless fundamental even in the major biogeochemical cycles, as I shall show. Microbes are *terraformers*: They have made the earth conducive to the development of many species, including humans.[8] Bacteriophage viruses act as "composters of worlds" and have unrivaled evolutionary powers (chapter 4). Under certain conditions, they can even cure humans made ill by pathogenic bacteria.

I posit the need to consider phage therapy not as an isolated practice but as one mode of relationships, *among others*. Using, exploiting, or repurposing some or all of these relationships requires us to pay particular attention to the ecologies in which they participate and in which humans are also involved, from the specific and occasional encounters that take place in laboratories or in patients' bodies, the recruitment of living beings for clinical trials, and the role of phages in the general process of carbon capture to the factory farms developed by capitalist societies beginning in the postwar period. These various scales cannot be assimilated: Each situation involving microbes and humans is specific. What we must consider, and what an ethnographic description following phages in these various *agencements* (a concept I borrow from Deleuze and Guattari)[9] can reveal, are the vital links between these scales and the scientific, political, social, and economic questions that they raise.

THE MANY RELATIONSHIPS BETWEEN HUMANS AND MICROBES

Viruses for healing. At first, this may seem like a counterintuitive idea. It may seem surprising that patients, microbiologists, infectious disease specialists, pharmacists, regulatory agency employees, patient representatives, and start-up companies are championing the development of phage therapy in the immediate wake of the COVID-19 pandemic. However, these developments must be seen in the light of the changes and reconfigurations of knowledge in microbiology and microbial ecology over the last twenty years. They form part of a scientific and medical context that both conditions the possibility of a renewal of phage therapy and imposes a set of constraints on the manner in which it could and should be implemented.

Since the beginning of the millennium, major changes have occurred in the approach to understanding microbes, based in particular on significant progress in genetic sequencing techniques and bioinformatics that have led to the development of metagenomics, that is, the sequencing of the genetic content of samples taken from complex environments such as fresh water from rivers and salt water from the oceans, as well as samples from the air, soil, and even human intestines, to name just a few examples. Metagenomics has led to a renewed appreciation of the importance of the microbiological composition of milieus, bringing to light a diversity that had previously only been suspected while also reopening and relaunching entire avenues of research, relating in particular to the consideration of interindividual, intercommunity, and interspecific (interspecies) dynamics.[10] Since then, the notions of microbiome (the entire genome of the microbial populations of an environment) and microbiota (the complex ecosystem of these microorganisms) have flourished, underlining the profoundly relational and systemic nature of the modes of existence of living entities. Microorganisms—bacteria, viruses, protists, amoebas, and fungi—are necessary for life to exist.[11]

These issues are also galvanizing human and social scientists, particularly those involved in *multispecies studies* who are analyzing the new attention being paid to these microbes in the natural sciences and in dietary practices and biomedicine.[12]

The anthropologists Heather Paxson and Stefan Helmreich refer to this change as the "microbial turn" to emphasize that these microbes are now a force to be reckoned with.[13] The field of biomedical sciences is permeable to these approaches, as shown by the recognition of the possible roles and functions of microorganisms that is driving the growing tendency to reconsider health statuses and the etiologies of numerous pathologies in terms of balances or ecological disturbances.[14] The geographer Jamie Lorimer describes human interventions as "probiotic" when they influence these balances, "using life to manage life, working with biological and geomorphic processes to deliver forms of human, environmental and even planetary health."[15]

Phage therapy could fit into this emerging analytical framework. However, the notion of "probiotic," forged in opposition to "antibiotic," tends to perpetuate the antagonisms ("pro-" versus "anti-") that assign living entities to one side or the other of an invisible line separating war from peace. In the logic behind this dichotomy, eradication seems to have been the preferred solution throughout the twentieth century and even beyond, as demonstrated by the martial rhetoric that accompanied the management of the COVID-19 pandemic.[16] To my knowledge, however, such a strategy has only really functioned on one occasion—in the case of smallpox—and it was mainly the characteristics of its pathogen that made this possible. However, this does not mean that the solution can be found by proceeding in a state of blind irenicism or that we should celebrate the "living world" without distinction. What microbes force us to consider is much more labile and uncertain. If everything is related, if species are heavily dependent on other species for their lives and survival, then clearly not all relationships are beneficial, and those that are beneficial at one time can just as easily become toxic at another.[17] In general, therefore, humans have had to *learn to live with microbes*, to domesticate them by implementing vaccination programs and preventive or prophylactic measures and treatments, the proposed responses to which have rarely been single and final.[18]

Recognizing the relational character of living beings should therefore not lead us to embrace an irenic conception but, on the contrary, enable us to perceive its *political* and *situated* dimension: In a given situation, which relations do some humans choose to make matter? Which relations do they consider insignificant or undesirable?[19] In addition, in line with decades of research on the history of human populations' interactions with pathogenic microorganisms, how do humans implement these relational choices?

ANTIMICROBIAL RESISTANCE AND THE INTRUSION OF GAIA

The philosopher of science Isabelle Stengers starts her book *In Catastrophic Times* with the disaster caused by Hurricane Katrina, which ravaged the Louisiana coast in August 2005, causing numerous fatalities, destroying the living environments of thousands of families and complex ecosystems, and fueling social and racial inequalities. In response to such disasters—all consequences of global warming—Stengers urges us to "name the event." She says that humans must address "the intrusion of Gaia":

> It is crucial to emphasize here that naming Gaia and characterizing the looming disasters as an intrusion arises from a pragmatic operation. To name is not to say what is true but to confer on what is named the power to make us feel and think in the mode that the name calls for. In this instance it is a matter of resisting the temptation to reduce what makes for an event, what calls us into question, to a simple "problem." But it is also to make the difference between the question that is imposed and the response to create exist. Naming Gaia as "the one who intrudes" is also to characterize her as blind to the damage she causes, in the manner of everything that intrudes. That is why the response to create is not a response to Gaia but a response as much to what provoked her intrusion as to its consequences.[20]

In this book, I consider bacterial antibiotic resistance as one of the manifestations of the intrusion of Gaia, as an impending disaster—a disaster that is already happening and in the making (chapters 7 and 8).

The rise of antimicrobial resistance is far from a simple public health "problem," an evil of the twenty-first century that could be resolved by the ingenuity of a handful of researchers or pharmaceutical companies and the discovery of new antibiotics or new molecules. The history of antibiotics—the history of how humans have repurposed molecules produced by microorganisms to eradicate living entities that are deemed undesirable—is invaluable in terms of the lessons we can learn from it and the matrix that it establishes, in which phage therapy is embedded. Looking back on this history, we can see that antibiotics have been much more than a therapy whose benefits to human health have been and remain unquestionable. In the second half of the twentieth century, because of their ease of production and administration but also the discovery of initially unsuspected functions, they formed a veritable infrastructure underpinning some of the developments of capitalism, by facilitating and thus exacerbating the exploitation of human and nonhuman living beings. This was a prerequisite for the intensification of production and consequently for the mass consumption of animals and plants (chapter 7). However, the eradicative logic that authorized these developments—generalized far beyond the primary objectives of caring for people suffering from bacterial infections—has had major consequences as it triggered changes in the quality of relations between living beings that have permanently transformed both human societies and microorganisms. The latter are sensitive to their environments, which are now saturated with antibiotics, and have adapted (chapter 8). Bacterial antibiotic resistance constitutes an event insofar as, once again, more than a century after Louis Pasteur's achievements, it raises the question of the relationships between humans and microbes. What it enables us to consider and what it brings to the forefront are the incredible evolutionary and adaptive capacities of living entities: This is the event that we must be able to name.

The idea behind this book is therefore not to rehash the apocalyptic prose cultivated in the reports of international bodies, which are unfortunately no more effective than those of the Intergovernmental Panel on Climate Change in eliciting responses, but rather to take sufficient time to describe and name situations that are always specific in a manner that encourages us to feel and think, to distinguish between "the question imposed and the question to be formulated," in the hope of triggering the creation of new collectives and new power relations. As we shall see, the question of antibiotic resistance is only one of the issues raised by chemical pollution. It is just as worrying and irreversible as the effects of climate change, albeit much less sensational, and requires a response.[21] Some practices involving phage therapy could inform experiments with other ways of orchestrating the relationships among humans, viruses, and bacteria. Using phages to treat antibiotic-resistant bacterial infections addresses the consequences of Gaia's intrusion. It may also, under certain conditions, inspire responses to the causes of this intrusion for the analysis of phage therapy practices shows that we can no longer contemplate prospective human interventions without considering the many relationships in which microbes are inextricably involved, as well as their adaptive capacities and agencies.

PRODUCTION, REPRODUCTION OF KNOWLEDGE, AND SOCIETIES: THE IMPORTANCE OF INFRASTRUCTURES

A therapeutic innovation is not in itself "good" or "bad." How it is designed and developed, and the consequences of its applications, will depend on how it accounts for the specificities of the human and nonhuman entities that it brings into contact, and the knowledge that it helps to create and on which it is based, as much as on the material, scientific, political, and social contexts in which it is used and that it will contribute to transforming.

The production of knowledge in laboratories and hospitals is unquestionably an essential step toward understanding the development of biomedical innovation (chapters 3 and 4). The subtle but significant nuances in conceptions of bacteriophage viruses and their modes of existence developed by scientists; the manner of studying, qualifying, and quantifying interactions between phages and bacteria and their subsequent use for treating sick people; the definition and recording of these people's health status (what is significant and what is not according to health care professionals, infectious disease specialists, or surgeons); and how these people's recovery after treatment will be established (chapters 5 and 6) are all conditions and propositions for ways of envisaging treatment and health that cannot be ignored if we are to succeed in accounting for what is at stake in the development of such an innovation, rather than limiting ourselves to a superficial commentary. For it is these norms and standards, and the moral economies they contain—patiently produced by trial and error, mobilizing specific disciplines, approaches, and methods—that phage therapy will need to bring into existence *outside* the spaces from which they emerged.

We have Bruno Latour to thank for the clearest explanation of what is required for the reproduction of experimental successes outside the laboratory, an explanation that is doubly compelling because it is based on the experiment conducted by Louis Pasteur in Pouilly-le-Fort in which he vaccinated an entire flock of sheep against the anthrax bacillus. Latour shows us that for the vaccine to exist outside the environment in which it was created, the farm on which it would be tested had to be transformed into an *extension of the laboratory*. The conditions under which vaccine efficacy was obtained in the laboratory had to be reproduced as closely as possible. The need to reproduce the conditions in which a scientific fact was initially produced to enable the fact itself to be reproduced will be of no surprise to scientists, who are accustomed to the materials and methods sections with which every article published in a specialized journal begins. These sections are intended to enable any other scientist in any other laboratory to perform the same experiment and thereby verify the published results.

What Latour encourages us to consider in his article are the transformations that are required *outside* the laboratory to reproduce and accommodate this new fact: "Scientific facts are like trains," he writes. "They do not work off their rails. You can extend the rails and connect them, but you cannot drive a locomotive through a field."[22] Phage therapy is no exception to this rule: It requires the production of phage collections, their storage, the implementation of treatment protocols, the training of health care staff, and the creation and maintenance of spaces. These are the main challenges and issues facing the living beings—human and nonhuman—involved in this work.

However, innovation never takes place in a vacuum, and, to use Latour's metaphor, it is a matter of understanding not only how the locomotive is constructed and how the rails are assembled but also how the topography of the land on which the tracks will be laid should be adapted. The history of phage therapy and its revival cannot be separated from the complex and tortuous history of antibiotic therapy and its corollary, antibiotic resistance. As I will demonstrate, the dominant model for the development of anti-infectives, designed primarily for chemical molecules, held back the development of phage therapy considerably while imposing a strategy of use that is poorly adapted to the specificities of phages. As an infrastructure, therefore, antibiotics hinder the development of a (partial) response to the problem they have contributed to creating.

More generally, and as shown by researchers in science and technology studies, the exploitation of the opportunities spawned by the knowledge produced in the life sciences in the second half of the twentieth century was indissociable from the rise of neoliberalism as a political force and an economic project. In particular, the sociologist Melinda Cooper shows how biological life—through recombinant DNA technology, the use of stem cells, and the practices of the pharmaceutical industry—has been transformed into added value with the creation of what can be termed "bio-economies" and various forms of biocapital.[23] The "biotechnological revolution" has therefore permanently reconfigured the relationships among scientific, economic, political, and social practices. In so doing, it has also renewed systems of domination and inflicted new forms of violence.[24]

How can we encourage the development of a therapy based on the potentialities of microbes while evading these deadly rationales? Scientists and doctors are working on alternatives that consider the specificities of phages and developing the moral and political propositions put forward by people exposed to the many relationships documented in this book to produce responses to the intrusion of Gaia that do not reproduce conditions that can lead to disaster. Such alternatives might seem virtually impossible, given the massive effort they require and the colossal social, political, regulatory, and economic transformations they entail. But these alternatives exist *thanks* to phages, thanks to their specific characteristics and their unique mode of existence. Because they are difficult to incorporate into some projects that humans may have in mind for them, and because they resist, they cause the actors who encounter them to hesitate and invent other possibilities.

EMBEDDED ANTHROPOLOGY AND SITUATED KNOWLEDGE

Hesitating. Making choices. The positions adopted in this book—mine and those of the people I have encountered—are snapshots stemming from knowledge that is always *situated*.[25] We are indeed at a point where nothing has been stabilized. Over the course of various encounters, dividing lines emerge and conflicting conceptions of care and health appear, relationships are forged and ended, and new knowledge is produced, bringing to light a particular entity, property, or capacity that was previously ignored or considered negligible. New beings appear, presenting novel or, on the contrary, all-too-clear issues.

As in the social sciences, no position is innocent in the life sciences.[26] There is no "view from nowhere." This book, in its design and structure, is the result of my encounter with not only phages and bacteria but also the people who develop these relationships. It therefore bears witness to how I was and became "embedded."[27] It reflects my amazement, but also my fears, in the light of the incredible potential capacities of these biological entities that defy classification. Phages can destroy bacteria. They do so with formidable efficiency as composters of worlds in perpetual action. However, they can *also* form temporary chimeras with them, accompany them into new spaces, and enable them to adapt to hostile environments saturated with antibiotics (chapter 4). What is more, bacteria can *also* learn to resist those phages that are seeking to destroy them.

A phage or viral particles? A bacterium or a colony of bacteria? Individuals or populations? Microbes could not care less about human scales and time frames or about the categories that scientists use to discuss them. In the following pages, and depending on the chapter, *phage* refers to either a virion or to all the viral particles produced after the encounter of a single phage with a population of bacteria. *Bacteria* refers to both a single cell and the billions of cells lining a Petri dish, which are considered identical. Rather than trying to specify what we might be dealing with on each occasion, I have preferred, with my interlocutors, to exploit these ambiguities—a way of reiterating that, as far as microbes are concerned, humans are in a permanent state of uncertainty. "Everything" is relational, but this "everything" is itself always interpreted in a partial manner, whether we are talking about the "everything" that my interlocutors are trying to comprehend or the "everything" that I wish to present in this book.[28] An "everything" that is ever-changing, imperceptibly or suddenly, in a state of permanent composition and recomposition.

We must contend with these ambivalences, and with an absence of innocence, including in nonhuman entities. And this is precisely what we must never forget, as I have learned from my many encounters over the past six years at conferences, congresses, and workshops; in microbiology laboratories, infectious disease departments, regulatory agency meetings, people's living rooms, and cafés; from scientific papers, reports, and guidelines published by national and European regulatory bodies and the World Health Organization; and from the press and books intended for a broader audience. All of the people involved in this work are seeking to promote phages as therapeutic entities for healing purposes; all of these people are producing knowledge as propositions according to their expertise, their discipline, their conception of care, or to more general principles. All of these people, or almost all, are hesitant.

In a 2012 article, the American anthropologist Kim Fortun called for ethnography to become an activist process. In response to the challenges of the period she calls "late industrialism"— characterized by degraded infrastructures, the inability of ill-suited paradigms to accommodate the complexity of current situations, and the primacy of neoliberal individualism over the collective—this is a matter of mobilizing the best of what ethnography can offer. The ethnographic approach can be used to describe situations and thus understand *what is*—discursively and materially—to account for various analytical perspectives and establish links between heterogeneous spatial and temporal scales. By mobilizing the knowledge produced by the description and presentation of actual collaborations with the interlocutors with whom we interact in the field, it can also generate new idioms and ways of thinking that embrace current realities and perhaps formulate precise responses to them. This is therefore a political approach "open to futures that cannot yet be imagined."[29]

Phage therapy is under development. Since nothing has yet been firmly established or completely decided, it is still possible to explore the forms that it could take. The challenge is to bring into existence descriptions, and therefore relationships, other than those that dominate the therapeutic field today. The nine chapters of this book present these many relationships, each proposing conceptions of what phage therapy is or should become. Examples include André, a paraplegic man who traveled to St. Petersburg to obtain phages that he could inject into his bladder himself to eliminate bacterial strains that had been "bothering him" for far too long. André, who considers it scandalous that phages legally produced and sold in Russia and Georgia cannot be used in France, raises questions about the regulatory existence of these viruses (chapter 1). Julie, meanwhile, describes her laboratory work not as the selection of phages that are particularly suited to "killing bacteria" but as the search for the best possible conditions of communication between these two types of entities. In this way, she and others are proposing an alternative approach to the dominant conceptualization of viruses as dangerous predators and paving the way toward the consideration of their incredible potentialities (chapter 3). In a departure from the central dictate in infectiology—of eradication being the main criterion for healing—Raphaëlle, an intensive-care physician who is regularly confronted with chronic bacterial infections, believes that phages can help patients live better with their infections (chapter 5). According to some scientists and physicians, phages, by targeting only the strains responsible for infecting people, may be able to protect bacterial populations whose role in our organisms or in the various ecosystems in which they participate is difficult to assess at the moment (chapters 5 and 6). We will also meet Belgian biologists who would like these phages to be accessible to anyone who needs them, at reasonable cost (chapter 9).

My aim is to describe and document these various positions as accurately as possible. However, no one is ever safe from accusations of betrayal by some or instrumentalization by others, especially in this story, which is as fluid and unpredictable as the entities that are its leading characters. I had to make choices about which avenues to explore. Despite the abundance of propositions and data generously shared by the people I encountered, I have chosen to concentrate on the use of phages in the field of human health, restricting mentions of other potential applications in animal health, the environment, and biocontrol procedures to clarifications concerning their use in humans. I have also focused on a particular geographical area: Europe, mainly France, Switzerland, and Belgium.[30] In their form, the propositions presented in this book are therefore specific to these countries and their infrastructures (especially in chapter 9).

I share a dual concern with many of my interlocutors: that of making treatments accessible—materially and geographically—to anyone who needs them while also establishing responses to antibiotic resistance that are as noninvasive as possible to limit disturbances caused to ecosystems and their microbiota, whose boundless diversity, assets, and functions are only just becoming apparent to scientists. The structuring of the descriptions, relationships, and propositions in this book meets both of these aspirations and is consistent with a reasoning that seeks to demonstrate why alternatives to the dominant model for the development of biomedical innovations are not only necessary but also, and specifically in the case of phage therapy, *possible*.

1

TENSIONS

I met André at his home in May 2018.[1] André, who has been paraplegic since a paragliding accident in the 1980s, suffers, like many paraplegics, from repeated urinary tract infections as the stagnation of urine in the bladder encourages the development of bacterial germs. Since his accident, he has read voraciously and performed extensive online research. He says, "When you're paraplegic, regardless of whether or not you're catheterized, you suffer from repeated urinary tract infections, and you therefore have germs that become increasingly resistant. So I'm quite careful and try not to take too many antibiotics." The loss of treatment efficacy is a constant in chronic infectious diseases: Germs adapt and become "increasingly resistant." Side effects, irrespective of whether they are listed in pharmaceutical package inserts, may develop over the long term and become so unbearable that there is sometimes no other solution than to stop the treatment causing them. André has obtained repeat requisitions for urine cultures, which determine the presence of bacterial species in the bladder. He has learned how to interpret the results; he knows the names of the various bacterial strains, their differences, and how to interpret an antibiogram: the microbiological test that determines the antibiotic sensitivity of the bacterial strains present in his bladder. André knows the treatments and the molecules' modes of action. At the beginning of our discussion, he therefore presents a detailed description of the various therapies—antibiotic or otherwise—that he has tried over the last thirty years, as well as their objectified or supposed effects.

MEDIATIONS

André first encountered bacteriophage viruses on a website several years ago but did not dwell on them, partly because his contacts in the medical world seemed to know nothing about these viruses and partly because the molecules used at that time were still sufficiently effective against his germs. It was only when the available solutions started to run out that he decided to delve deeper into what seemed to be an alternative solution and try it. However, phages were not available in France or in most European Union countries. The first challenge was to obtain some:

> It always boils down to Russia, you know: Georgia, Russia. As soon as you read something [about phages], they tell you it's in Georgia or Russia. Georgia seemed very remote to me, but my mother had been to St. Petersburg and came back delighted. She had told me that it was a magnificent country, and I thought that it would be a good idea to take a look at what the Russians had to offer. Finally we went on a week-long trip with one of my daughters. And I thought I'd have a look to see whether we could get hold of these phages because there was conflicting information about whether they were sold over the counter. In fact, nothing was very clear.

On the last day of his stay in Russia, André decided to venture
into a pharmacy:

> Since I didn't speak Russian, I had jotted down "bacteriophage
> *Pseudomonas aeruginosa*," which was the name of the germ that
> was bothering me, along with the name of another germ. The
> pharmacist looked at me and said, "OK." He left and came back
> with two different boxes. The labels were in Russian, but the
> names of the bacteria were noted in Latin characters. I took both
> boxes, put them in my luggage, and we returned to France. And
> then I said to myself, "What on Earth am I doing?" I'd willingly
> try to heal myself with this, but what if something should hap-
> pen to me? I had bought them in a pharmacy; I certainly wasn't
> afraid of taking something dangerous—that wasn't the problem
> at all . . . What's more, it all seemed very serious; you look at the
> packaging—it's a medicinal product. But it was more a question
> of what will I do if there really is some kind of effect, an allergy,
> or whatever? Should I be doing this all on my own? In addition,
> we also had to look for instructions in French. But I solved all
> these problems. However, how should I go about putting them
> in my bladder? It would surely be better if it were in a medical
> environment—in a hospital. So we looked into it.

André decided to consult his infectious disease specialist at a
hospital in the suburbs of Paris, who had been recommended
to him when he first started talking about phage therapy. Dr.
Prime, "a great fan" of phages according to André, agreed to help
him with the treatment, as did Dr. Alessandrini, a physician in a
large Paris hospital, with whom he was already in contact:

> I was delighted. We had practically made an appointment; every-
> thing was set. I had put both doctors in contact with each other, so

everything was synchronized, and both seemed very happy to be collaborating. And then something happened; I don't know what. They looked into it; I don't know what they did. Dr. Prime put a stop to everything by sending me an email as if we had barely mentioned the case, more or less stating that he didn't want to get involved in this matter—well, in slightly different terms—but I felt that he was sending me this email to cover himself. And an embarrassed Dr. Alessandrini also told me that we couldn't proceed, whereas she had actually said yes. Everything had been planned with Dr. Alessandrini: that I would have such-and-such an examination beforehand, that I would be hospitalized. . . . In short, everything was in place, and we just had to choose when to do it. Well, I wasn't very happy, and so I started to think something was wrong: There was a drug; you couldn't test it, but apparently it works everywhere. I was starting to become not a specialist but to understand the subject nonetheless—what on Earth was going on?

In spite of her refusal to take medical responsibility for administering the phages, Dr. Alessandrini agreed to carry out in vitro testing of the medicinal products on André's germs, in particular on "[his] *Pseudomonas*," the problem that was clearly bothering him the most. She first asked him to provide a urine sample, as well as a sample of the phages brought back from Russia to verify their efficacy, and then confirmed that both products worked, one better than the other:

I wanted to get rid of my germ, and so I said to myself, "What the heck, I'll give it a go." But I want to do it right; I'll try it and see what happens. However, I had discovered that there were many methods of administration, even though according to what I had read, you could take everything by mouth. That is, if you had a foot infection, you took it orally; if you had a urinary tract infection, you took it orally; it seemed a little weird to me. I had also read that you could wash a wound with it. So, once again, this got me thinking, "What on Earth am I doing?" Should I try to inject it into my bladder? If so, I'll need a catheter because I want to perform multiple lavages. How am I going to get hold of all this? I'll need a syringe for sterile injections into the catheter. Then came a first pleasant surprise: At the pharmacy, they'll give you a syringe without asking any questions. I could purchase sterile syringes for ten cents. It was trickier to obtain a catheter without a prescription because I couldn't ask my doctor. But I discovered that you could order samples, so I called all the labs that could provide me with samples—ten catheters at a time, which was sufficient. And then I started. But I had a urine culture analyzed just before— I was trying to do things right! I waited for the results to make sure that the germs were present, even though I had all the symptoms, which had been persisting for a long time. I could tell that my condition had stabilized, but I wasn't doing great; there was a germ that was bothering me. And I thought it was *Pseudomonas*. I was also afraid of *Pseudomonas* because my mom was a nurse who had been in the Vietnam War, and she was on the ships that brought the casualties home. And when she found out that I was a paraplegic, she envisaged me dying of a urinary tract infection caused by *Pseudomonas*. And I recall the first time it showed up in my urine culture before she passed away—it was a long time ago. When she saw that I had *Pseudomonas*, she was not happy at all. . . . I started [administering phages] on a Sunday, I think. That's when I developed symptoms that I had read were common when taking phages: I felt poorly during the night, not really feverish but unwell; I slept badly. I was thinking, "What is this thing?" I was wondering, "Should I stop it or carry on? Should I continue? I'm not all that sick. They mention the possibility [of] this kind of symptoms on the websites. I'll carry on." So, I kept on with the treatment for a little longer by trying to do a daily injection into the bladder and then drinking as well. I took two different medicinal products. The one that was apparently more effective according to the package insert, and also according to Alessandrini's analysis, was X. So it was X that I needed to take. But I said to myself, "It's better to make a real impact, so I'll take both anyway." And so I also drank the Y, and then I alternated for a while like that. I may also have injected the Y into the bladder once; I don't remember very clearly. Anyway, I had that one bad night, but it soon started to improve. It became very, very clear that it was getting better, like when you take antibiotics; but with antibiotics, it takes a little longer to happen, two or three days. That's when I felt *phew!*, which convinced me to continue. On Friday there was nothing left; I had finished the treatment. And I felt much better; I was cured."

André wanted to objectify this result and so compared the cytobacteriological analyses of urine that he had conducted at different times:

Well, for the first urine culture, there was *Pseudomonas*, and there was also an *Enterococcus faecalis*, to which I had not paid attention, and *phew*, that had disappeared. So this *Enterococcus* had disappeared, but I said to myself, "I'll just see how long I had it." In fact, I'd had it for a long time, but I was focusing exclusively on *Pseudomonas*. And this *Enterococcus* had persisted throughout treatments with cipro[floxacin]—I'd followed several treatments—which also worried me. So I had this *Pseudomonas*, which had been dragging on for a long time, which had made me quite poorly and which I was pretty keen on getting rid of but which still allowed me to live with it. But I'd always told myself, "The day you take cipro, that'll wipe it out." But that's not what happened.

André then explains to me how he was prescribed the antibiotic ciprofloxacin several times but always, he believes, in insufficient doses:

> Because when you're used to it—and now I'm starting to understand how it works—when you really want to treat a germ in a paraplegic, if you really want to eliminate it, you have to pull out all the stops: You need a good fifteen days of high-strength Cipro. You mustn't go easy on it. But each time, they went easy on it. And I was wondering, "Well, if this thing survives, it'll get a little bit tougher each time." And indeed, it survived. What I discovered was that the Pseudomonas had survived, but so had this *Enterococcus*. But while the *Pseudomonas* had previously been considered sensitive [to ciprofloxacin], now it was resistant! That's it! That's what happened! The treatments were quite feeble; that's why it survived. It could have probably been eliminated at the outset, but then it became established. So, anyway, when I took the bacteriophage treatment, it eliminated the *Enterococcus*, which had also survived all those treatments. All in one day! And then I got better. So this made me realize that it's not the *Pseudomonas* that's harming me. When I'm sick, there's an additional germ, and it is this additional germ that we must try to treat. For example, at the moment, I've had several examinations since the treatment, and in February I started suffering from something again. I looked: There's a *Klebsiella*, which was not there before; I'd never had a *Klebsiella*! It appeared out of the blue, and now it was bothering me. Well, at least it's not resistant! But I didn't want it to be made resistant by carelessness. There you have it. All that to tell you that, in my opinion, phage therapy works like a charm. It's not right that it's not authorized, and I have lots of, how shall I put it, opinions on this subject.

André has been living with infections—a term that he does not use often—for more than thirty years. His understanding and apprehension of a general phenomenon—infection—is indeed above all relational: It is the result of an interaction with one or more organisms that he has learned to name and differentiate, such as bacteria of the species *Pseudomonas aeruginosa*, *Enterococcus faecalis*, and *Klebsiella* (figure 1.1).

For a long time, André had access to solutions that enabled him to "live with" his Pseudo. In contrast to certain discourses on the living world, however, and to an acceptance of living with these problems which might equate to a form of peaceful, or even irenic, coexistence, his experience of living with these germs was not devoid of conflicts or power struggles with them. These are entities whose characteristics are never intrinsic but rather defined entirely within and by relationships: Bacteria are "strong" or "weak," "sensitive" or "resistant"; actions carried out in the present or in the future are expressed in varying degrees of intensity. Each of the various conditions—physiological and psychological—that affect André is an indication of the strong or weak presence of undesirable germs, which are sometimes difficult to identify.[2]

In this type of situation, antibiotics, often presented as a radical solution to bacterial problems, are themselves caught up in these power struggles: Ciprofloxacin should have "wiped out" the *Pseudo* that was bothering André so much. But it had been misused, administered in doses that were far too small. In response to these "somewhat feeble" treatments, the Pseudo, which was sensitive to the antibiotic, was able to adapt and become resistant. This does not mean that André dreams of drowning his bacteria in a flood of chemical molecules. On the contrary: "I'm quite careful and try to avoid taking too many antibiotics." For example, when tackling the *Klebsiella*, a new entity for André,

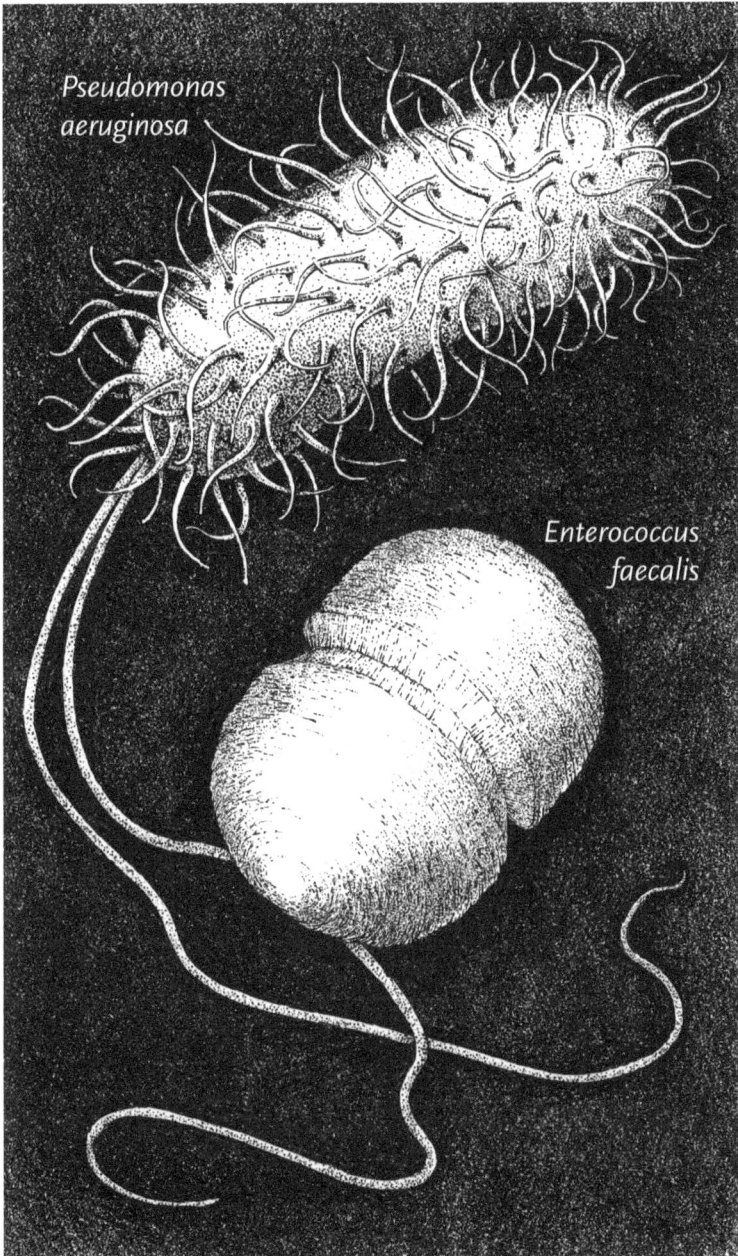

FIGURE 1.1 *Pseudomonas aeruginosa* and *Enterococcus faecalis*

it was important not to "make it resistant by carelessness." For André, what counted was making no mistakes, not *adopting the wrong approach* to the bacteria, for example, by failing to consider the adaptive abilities of this microorganism or forgetting that it is a living being that acts and reacts according to its own time frames.[3] Antibiotics are chemical molecules, most of which have a broad spectrum of action and are capable of killing various bacterial species. But they are also capable of making bacteria resistant if the bacteria have the time and resources to adapt to their new environment (i.e., that with the antibiotic present). (I will discuss these aspects in more detail in chapters 5 and 8). André's apprehension, perceptible throughout the interview, relates precisely to this possibility of carelessness over which he has no real control because once a bacterium is resistant to a given molecule, the alternative therapies available to him are drastically diminished. And the prospect of suffering, of burning pain, of being in permanent discomfort, of sleepless nights, then painfully emerges.

We must situate André's search for alternative treatments within the perspective of power struggles with an always uncertain outcome because mistakes will always be made, because bacteria are living beings, and because the germ that makes us suffer may not be the one we were expecting. Thirty years of living with the awareness that his health depends partly on the microbes that populate him and how they are treated is what André shares with me in this interview. It is an intimate, embodied knowledge of the bacteria that are intermittently present in his bladder. A story in which he translates his physical and physiological sensations precisely into so many traces and clues capable of accounting for this microbial life, his "living with" microbes. The difficulty, however, resides in the actions of others, which could make his *Klebsiella* resistant—because André cannot tackle his microbes alone. These relationships are mediated by chemical molecules that cannot accomplish everything but also by doctors who do not know everything, by limited mechanisms, by instruments to be perfected, and by forgotten or ignored therapies. "Phage therapy works like a charm!" André says. But success cannot be sought in the eradication of these germs; rather, it must emerge from the keys to understanding—and action—that it offers this man, providing one more thread, an unexpected and salutary mediation that could help him find the correct response to the germs that he carries.

ANGER

André's anger, which he vents in the second half of our interview, should be understood in light of the opportunity presented to him because to describe him as angry would be an understatement. The fact that he makes no effort to conceal his anger is what gave me the opportunity to meet him. Elisabeth, an employee of France's National Agency for the Safety of Medicines and Health Products (ANSM), created this opportunity during one of our meetings a few weeks earlier. As I was telling her about my desire to meet people who had received phage therapy, she started telling me about a man who had been denied this treatment. She mentioned having had some difficult discussions with him over the previous months, a conflict that I could only partly understand because Elisabeth could not give me too much information without André's consent. But she soon obtained this because André wanted to tell his story, to raise awareness of his experience, his suffering, his efforts to obtain recognition for a therapy that he was convinced contained the seeds of his deliverance. This is what he did that afternoon in early May. Given the sudden refusal of his two doctors to support him in his desire to pursue phage therapy, he was indeed "not very happy," a phrase that stands out as an understatement a few years later, to put it mildly. André was then, and continues to be, outraged by a situation that he finds incomprehensible. The root of the problem is not the mistakes committed in treating his bacteria by administering antibiotics at doses that were too low (or too high, depending on the case). In our interview, he is angry because he traveled to Russia; he purchased phages in a pharmacy; he asked for support; he was told yes, then no; he defined his own protocol; he obtained syringes and catheters; he administered the phages directly into his bladder; and he monitored himself by conducting cytobacteriological analyses of his urine at regular intervals. His *Pseudo* is still there but has diminished, and his *Enterococcus* has disappeared.

André is all the more angry because he has now developed all the zeal of a convert. He has expressed his anger in numerous letters to the ANSM; to Brigitte Macron, the wife of Emmanuel Macron, the president of France; and to Macron himself. Letters describing what he calls a "health scandal." During the second part of our discussion, as he is telling me about all the strategies he is thinking of adopting to "expose the scandal," some of which he has already implemented, he regularly returns to the deaths that could be prevented by phage therapy, not to mention the amputations. "Ten thousand deaths!" he repeats regularly with an ardor and rage that intensify with the increasing awareness of his impotence, sometimes jabbing his index finger into the table. "Ten thousand deaths!" Each time, I waver between shock and an irrepressible urge to laugh, the kind of nervous laugh that arises when you are caught up in a situation whose violence is real but partly escapes you. A nervous laugh, not a snigger; this is an important difference because it would be very easy to undermine André's discourse, to discredit him for his outrageousness and his tendency to evoke a "conspiracy," to exclude André from the discussions because of his propensity to disregard the implicit rules of reasonable discussion. What fascinates me is how he interprets his life *politically* through microbes: André has a problem with certain microorganisms; there seems to be a solution, but this solution is not available in France, and when he goes abroad to obtain it, nobody wants to help him with its administration.

"Phage therapy works like a charm! It's not right that it's not authorized, and I have lots of, shall we say, opinions on this subject." Opinions that he has been keen to share, first with Elisabeth in an attempt to understand why the ANSM did not want to support him and why, in his view, the agency encouraged Drs. Prime and Alessandrini to withhold their assistance. Opinions that he then sets out to share with other patients by creating a website and thereafter with political leaders and people identified as experts. For example, he emailed all the members of the second Temporary Specialized Scientific Committee (CSST) on phage therapy, organized by the ANSM, the night before the committee met on March 21, 2019, to inform them of the manifest lack of public desire to develop this therapy, attaching copies of the letters he had sent to various political leaders.

The reactions to this email were revealing as they conveyed some of the tensions associated with the development of phage therapy. The CSST included researchers (myself included), along with representatives of health care professionals, the Directorate General of Health, the Directorate General of Health Care Provision, and patient associations. Its objectives were to review the progress of phage therapy in France and the number of phages available, as well as to facilitate the provision of feedback on people who had been treated with phage therapy.

(I will describe the work of this committee in detail in chapter 9).[4] Although André's message was not mentioned during the sessions, it did generate some discussion during breaks. Some attendees were annoyed by his initiative, while others simply rolled their eyes. Indeed, André was known by many of those present at the meeting, and he aroused contrasting feelings. For example, Alain-Michel Ceretti, then the president of France Assos Santé (an inter-associative organization created in March 2017, whose mission is to represent patients and users of the healthcare system in France), whom I had interviewed in 2018, described his sympathy for André and the efforts he was making in an attempt to obtain recognition for phage therapy. He essentially told me that André's anger is not unjustified and that it needs to be heard—all the more so as, to a certain extent, it was shared by the patient associations represented that day, as well as by many people who have had an opportunity to encounter these viruses.

These include patients suffering from chronic, sometimes incapacitating, infections, who have conducted online research or been contacted by relatives who have seen a television program on the subject, following which they sought out phage therapy but were stymied on discovering that a therapy whose efficacy is frequently claimed to be proven in Georgia and Russia is not available in their own country. This was the experience of Christophe Novou-dit-Picot, whose story was heard during the CSST meeting. After several decades of suffering from treatment-resistant bacterial infections, he sought treatment in Georgia, thereby avoiding the promised disarticulation (total removal of a limb) that awaited him. Since then, he has created the "Phages sans frontières" ("Phages Without Borders") association to support people—several dozen now—who wish to undergo phage treatment by directing them toward the most suitable establishments in Georgia and by recommending interpreters to translate from French into Georgian. He claims that the association is not intended to last but rather to fill a void pending "the structuring of extensive access to phage therapy at the national level in response to the needs and thus avoid the perpetuation of this type of circuit."[5]

Novou-dit-Picot's position is shared by the medical internist Jérôme Larché, who founded a similar organization, "Phagespoirs," after accompanying his brother, who has cystic fibrosis, on a trip to obtain treatment in Georgia. People with cystic fibrosis develop many lung infections that arise from the environment created by the secretions affecting their bronchi. Recurrent antibiotic treatments often render the bacteria they carry resistant to antibiotics. Cystic fibrosis is therefore one of the conditions for which phage therapy could be considered. Larché's brother learned about phages by watching a documentary. He sought advice from his brother, who took a few months to review the scientific literature, assess the current situation, and evaluate the relevance of phage therapy for his brother's condition. Larché's brother returned from Georgia with aerosols containing phages and a greatly improved physical condition. Since then, Larché has been avidly monitoring the developments in phage therapy. He explains that while it is clearly not a miracle cure and should not be presented as such, it deserves to be made available in France; however, this would require a strong public will and the production of sound knowledge.

At the CSST meeting, I also encounter the president of a patient association who angrily deplores the recourse to health care facilities in Georgia. She is concerned about the information circulating on social media platforms, which she sees as providing excessive encouragement to seek treatment in Georgia when current knowledge seems to fall well short of meeting the necessary evidentiary requirements. She is indignant about a communication bias in which patients who have been cured are more likely to communicate about their treatment than are those who return to France still sick. This woman considers André's demands to use Russian or Georgian phages on French soil unreasonable. And dangerous—not so much because of the possible risks and side effects of treatment but because of the dashed hopes people would experience in the event of failure and the delays that such trials would incur in undergoing treatments already available to them at home.

In May 2018, after an interview lasting just over three hours, I left André's home with the certainty of having obtained a form of total narrative, that of a man engaged in politically interpreting his experience of living with microbes while presenting all of the data relating to a particularly complex problem. At the very least, this was as a story of a man of convictions, a man who went so far as to obtain viruses from St. Petersburg to inject into his bladder via catheter in an attempt to eradicate his body of microbes causing him significant distress, while monitoring his condition by conducting cytobacteriological analyses of his urine. This was "do-it-yourself" (DIY) phage therapy—the pursuit of which was a decision *driven by spite rather than choice* for although André makes political demands, they differ from those often found in DIY movements; namely, empowerment and the reappropriation of knowledge.[6] It is not a lack of knowledge about phage therapy in the French health care system that fuels his anger. He is acutely aware of the ignorance of nursing staff because he has been exposed to it on many occasions, yet he feels no animosity toward it. His anger stems from the fact that he is aware of the availability of a therapy in one country *and* of the impossibility of validating its use in another—an impossibility manifested by the doctors' refusal to help him when he returned from Russia.

André's anger resonates and is reflected in the positions of the actors present at the CSST meeting and beyond, within an ever-growing community. It is an anger that cannot be dissociated from the context in which these discussions are held. I was put in touch with André by an employee of the ANSM, a central authority omnipresent in the issue at hand because it ultimately draws the categorical dividing line between what is considered a substance (chemical or biological) and what is considered a medicinal product (i.e., a substance formally acknowledged to have the ability to treat a medical condition). This dividing line is situated in and fluctuates according to era and to the types of molecules or diseases being considered.[7] It is the ANSM that has the power to sanction a change in the ontological status of a given entity and make it a recognized therapeutic substance; it is the ANSM that can decide to send phages back to research laboratories or give them a new life in hospital pharmacies, startups, or pharmaceutical companies. Elisabeth did not point me toward André without reason; I am convinced that she expected me, a researcher at the French National Center for Scientific Research without any links to the ANSM, to convince André that he was mistaken, that there was no "scandal" to be revealed. André's letter to all CSST members one year later reveals that I failed to meet not only Elizabeth's expectations but also—and this is of greater interest to me at this point—André's. This is a failure that he would mention in a letter sent a few months after our interview to a member of the parliamentary group of la République en Marche (Emmanuel Macron's political party): "Even Charlotte Brives, a social scientist, failed to see the extent of the scandal."

So, how can we make manifest the problem posed by André's anger? How can we do him justice without interpreting the situation as a scandal? André's rich narrative gives us various avenues to explore: a forgotten therapy, an opportunity to precisely mediate his relationship with bacteria causing infections that need to be treated, people and microbes that cross borders in an attempt to put an end to suffering that has persisted for too long, doctors who are ignorant versus others who want to help, phages that exist but about which little is known in terms of their administration. In André's narrative, complex threads are woven and knots created regarding the production of knowledge, the administration of evidence, and the drug market. Taking André's anger seriously means taking the time to undo and reweave all those threads, starting with the first: the return of a forgotten therapy.

2

ALTERNATIVE HISTORIES

THE RISE AND FALL OF PHAGE THERAPY

"Georgia, Russia. As soon as you read something [about phages], they tell you it's in Georgia or Russia."[1] The reason that phages are found in these countries can be found in the history of the development of phage therapy, which I shall now touch on to describe some of the milestones that led to the "rediscovery" of this therapy today.[2]

Bacteriophage viruses were discovered in the context of certain activities from which the field of virology would gradually emerge. Two names are regularly associated with this discovery in a dispute about anteriority that persists to this day: Frederick Twort (1877–1950), of England, and Félix d'Hérelle (1873–1949), of France. In 1915, Twort published an article in which he assigned certain characteristics to a nameless entity that, as we shall learn in the remainder of this book, belong to phages. In 1917, d'Hérelle published an article titled "Sur un microbe invisible antagoniste des bacilles dysentériques" ("About an Invisible Microbe Antagonistic to Dysentery Bacilli") in which the term *bacteriophage* appeared for the first time. Although d'Hérelle made no mention of viruses at the time and had no means of observing them (the viral status of bacteriophages would be confirmed, by electron microscopy, only in 1941), he nevertheless deduced from his various experiments that they were "living germs" that destroyed bacterial cells. Over the following years, he was involved in the development of therapeutic applications, that is, the use of bacteriophages to treat people with bacterial infections, a practice that would soon be discussed in scientific journals, the media, and even literature.[3]

In the 1920s and 1930s, clinical trials were launched in many countries for a wide variety of diseases according to highly varied therapeutic regimens and theories, for example, dysentery in Brazil in 1923, the plague in Egypt in the 1920s, and plague and cholera in India in the 1920s and 1930s.[4] At the same time, phage-based preparations were being produced by university laboratories and pharmaceutical companies. In the 1930s and 1940s, bacteriophage-based products were marketed in France, Great Britain, Germany, Italy, and the United States, with disparate production methods and quality. In 1927 in France, d'Hérelle established the Laboratoire du bactériophage, which experienced many highs and lows until its closure in the 1970s. The products marketed included Bacté-coli-phage, Bacté-intesti-phage, Bacté-dysentéri-phage, Bacté-pyo-phage and Bacté-rhino-phage, which were also distributed in England and Germany. In the United States, several companies, including Eli Lilly, Squibb & Sons, Abbott, Parke-Davis, and Swan-Myers, also embarked on this adventure. In Germany, phages were available in tablet form as of 1927.[5] In Brazil, the Raul Leite Laboratory produced ampoules of Estafilofagina and Colifagina. Phages were also widely used in the Belgian colonies to tackle dysentery in 1943 and 1944 and to combat typhoid fever in Italy and Greece.[6]

However, the 1940s marked a decline in phage therapy. The manner and rate of this decline were specific to each country, and little attention has yet been paid to the complex reasons for it. Nevertheless, several explanations have been proposed. In particular, various controversies surrounding the modes of existence of phages (enzymes or germs), in addition to their modes of action and therefore the manner in which they could be used, prevented the necessary homogenization—and above all the standardization—of therapeutic practices.[7] These difficulties were compounded by inconsistencies in the production of phage preparations.

These production problems either did not affect or only briefly affected the chemical molecules that were also being developed at this time. Sulfonamides—powerful antibacterial agents developed in Germany—alone may not have been able to eclipse bacteriophage viruses, especially because of the serious side effects they could cause. Antibiotics, however, played a major role in the relegation of phages, especially since the conditions for their mass production were in place by the early 1940s. (I will revisit the history of antibiotics in detail in chapter 7). With penicillin being widely used in US hospitals by 1944, the therapeutic use of phages and the corresponding scientific literature declined significantly in Western Europe and the Americas.

However, it would be wrong to state that phage therapy simply disappeared during this period. First, not all bacterial infections—dysentery, first and foremost—could be treated by penicillin (which is still the case today, despite the large number of antibiotic molecules available, as I will discuss in chapter 5). This led some scientists, such as the eminent microbiologist René Dubos, to undertake research on the efficacy of phages to ensure the availability of reliable alternative therapies, especially after learning about the results obtained by Soviet scientists, as we shall see. Second, doctors soon became aware of the adaptability of bacteria to antibiotics and the risk of antibiotic resistance. For this reason, some of them tested strategies that used phages and antibiotics in tandem.[8]

There is still a need for extensive historical research on the therapeutic use of phages from the 1940s to the 1990s, a period delimited by the establishment of antibiotic therapy, on the one hand, and the development of bacterial antibiotic resistance as a public health problem, on the other. Nonetheless, precious details were provided by a medical thesis defended by Gaëlle Bourgeois at the Hospices civils de Lyon in 2020. In her thesis, Bourgeois draws on the medical archives at her disposal to show that three distinct periods can be identified with regard to Lyon specifically and to France in general: a period until the 1940s during which phages were used primarily to treat dermatological diseases (e.g., boils) with excellent results; a period of decline during the 1940s and 1950s, undoubtedly associated with the addition of antibiotics to the anti-infectious therapeutic arsenal; and a period that saw the resumption of phage therapy from the late 1950s until the 1970s to treat infections resistant to the antibiotics available at the time. However, while phage therapy might otherwise have enjoyed a new lease on life in this latter period, its use ultimately petered out in the 1970s and 1980s. Bourgeois outlines important hypotheses for the disappearance of bacteriophages in the Lyon region at the end of the 1970s, showing how knowledge of and expertise with these viruses had gradually been lost in the preceding years, notably owing to a lack of resources and the need to tackle other diseases such as poliomyelitis from the 1960s onward.[9]

In France, the 1970s were marked by the closure of the Laboratoire du bactériophage in Paris and a break in the transmission of knowledge and expertise. Neither of the two public laboratories that produced phages in France until the early 1980s—one in Paris and the other in Lyon—survived the retirement of its director. Phages were removed from the *Vidal* (the French pharmaceutical dictionary) at the end of the 1970s.[10] As for the few remaining vials, they have been preserved, sometimes exhibited proudly as vestiges of a bygone era, and now, when it comes to exhuming and reviving phage therapy, serve as a tangible link to this long tradition.

However, there is another, more clandestine history of phages. Long after the last laboratories closed, doctors continued to treat certain patients with phages that they obtained from countries of the former Eastern bloc. When I decided to look into phages in 2016, I was introduced to this history by one of its main protagonists: Alain Dublanchet, a now-retired infectious disease specialist who worked at the Villeneuve-Saint-Georges hospital for many years. One afternoon, he told me about his encounter with phages and how, primarily in the 1990s and the first decade of the twenty-first century, he proposed to use bacteriophage viruses to treat people who had experienced therapeutic failure.[11]

AN ALTERNATIVE HISTORY

Not all therapeutic phages have been left to sediment in dusty vials. Although willingness to use phages in Western Europe and North America today appears to be a relatively recent phenomenon, the practice has been part of the continuum of health care provision offered by certain former Soviet bloc countries since the late 1920s. The origin of what could be considered an alternative history can be traced back to a meeting between Félix d'Hérelle and Giorgi Eliava, a Georgian physician and biologist, that took place at the Pasteur Institute in Paris at the end of the 1910s. The relationship forged by these men over the following years led to the establishment in 1923 of a center for the study of bacteriophages, which will become known as the Eliava Institute". D'Hérelle also made a substantial contribution to the development of other sites in Kyiv and Kharkov.[12]

Little research has been done on phages, phage therapy, and the persistence of phage therapy in countries of the former Eastern bloc. According to Dmitriy Myelnikov, who has studied the archives of the Eliava Institute, three factors account for the continued development of phage therapy in these countries, even after 1950 and the introduction of the means to mass-produce penicillin.[13] First, research on bacteriophages and their practical applications was not only consistent with the ecological conception of bacterial infections that predominated in Soviet microbiology between the two world wars but also prevailed in the very economics of Soviet research institutes, which carried out research, clinical trials, and the production of therapies.[14] Second, the creation of the Tbilisi Institute in 1935 perpetuated this system in Georgia. The progression of knowledge and technical expertise was subsequently illustrated during the Winter War between the Soviet Union and Finland (1939–1940) and during the Second World War, when phage-based preparations were used to combat dysentery and gangrene. This success was acknowledged by both doctors and military authorities (and, beyond that, by microbiologists such as René Dubos in the United States).[15] Finally, Myelnikov considers that the extreme isolation of Soviet scientists during the Cold War protected the practice of phage therapy from the criticisms that it encountered in the West. In this way, it remained an alternative to antibiotics or a complementary treatment. Although phage research declined throughout the Eastern bloc during the Cold War, as in Western Europe and North America, Georgian scientists have managed to maintain close links between fundamental research and therapeutic applications to this day, making this story "an alternative to the familiar narrative portraying the dominance of chemical medicinal products."[16]

However, as outlined in chapter 1, there seems to be a certain mistrust of Georgian knowledge and expertise, which is at odds with the enthusiasm of people who have been to Georgia, those who plan to go after learning about the availability of phages there, and those who have read testimonials or reviewed discussions on the various websites and social media accounts specializing in this matter. A substantial gap has grown between a large proportion of the scientific and health care community and what are commonly referred to as health care service users.

I first noticed this gap at a symposium commemorating the hundredth anniversary of the publication of Félix d'Hérelle's paper on bacteriophage viruses held at the Pasteur Institute in 2017. While I was reading the posters prepared by the speakers during a coffee break, a researcher working in the United States who had discovered that I was an anthropologist approached me with a very specific question: Why, in my opinion, did Georgian scientists spend so much time claiming that they knew everything about phages and their therapeutic uses while simultaneously refusing to share this knowledge with the rest of the scientific community? At first, I was a somewhat taken aback by this rather violent charge against researchers who had presented several of their studies during the colloquium sessions and who had also produced some of the posters presented before us. But I was particularly annoyed by what I construed as an outright dismissal of Georgian knowledge and practices—a dismissal that was clearly perceptible in the tone and formulation used. Asking the question in this way implied that the Eliava Institute's staff were engaged in little more than window dressing. In other words, if the Georgian scientists indeed knew everything, there would be no honest reason not to share their knowledge. I was torn between two responses to this researcher's question. My first instinct was to consider the institute's dismal and chaotic fate, which I had just read about in Anna Kuchment's *The For-gotten Cure*, a story of dispossession, unequal partnerships, and plundering by North American start-ups that had bled the institute dry. A story that, in itself, could have been sufficient to explain Georgian scientists' wariness toward the demands of their European and North American counterparts. But I was also contemplating the consequences of decades of anticommu-nist propaganda designed to disqualify the knowledge produced in the Eastern bloc, from which the people I encountered during my investigation were not always immune. Beyond this bipolar-ity, reminiscent of the heights of the Cold War, my interlocutor's attitude, however caricatural it may have been, merely summa-rized reactions and arguments that I would regularly observe over the following years. Either the Georgians were lying about the extent of their knowledge, or their knowledge was valid but they were keeping it to themselves for somewhat dubious reasons, mainly financial. Or it was a mixture of the two, the omission serving to conceal both what the Georgians knew and wanted to keep to themselves *and* what they did not know and did not want anyone else to realize they did not know. How-ever, the uncertainty of the situation does not prevent most sci-entists and physicians from respecting the work accomplished at Eliava. The mistrust is generally subtle, relating to scientific and technical aspects such as the question of method of admin-istration, which crops up time and again. Do "the Georgians" typically administer phages orally?[17] *But* that would not be effec-tive: The pH of the stomach is too acidic, and the phages would not survive there. Do they administer them as a local application in topical form (e.g., in a cream)? *But* that would work only if the targeted bacteria were on the surface of the skin. As far as evaluating efficacy, "the Georgians" claim that patients are cured of their infections. *But* how can they be sure that it is because of the phages? Were the patients cured upon their return from Georgia? *But* what if they had already been cured before going there? The same applies to production methods: "The Georgians" say that they make their phage preparations in vats. *But* how can we know what is in them? They say that their phages are active and of high quality. *But* what bacterial strains do they use to produce them?

There is always a *but*. "They say they know, *but*. . . ." "They say it works this way, *but*. . . ." "They say they cure the sick, *but*. . . ." "They say they can produce phages like that, *but*. . . ." After a while, this succession of *buts* can be seen in its true light: Not only does it convey Western researchers' doubts about the quality of Georgian research, but it also, and above all, draws attention to, or even exorcises, the scientific community's relative ignorance of all these issues. The operating mechanisms, evaluation of efficacy, and even production standards are all ongoing projects for which nothing has yet been stabilized or standardized, and the number of cases treated in Europe in recent years has been far from sufficient to create a consensus on these questions. However, although the limitations of available knowledge and expertise partly explain the gradual abandonment of phage therapy in the West in the 1940s, the existence of Eliava and, more generally, of phage-based knowledge and products in countries of the former Eastern bloc serve as a painful reminder of what has been lost and forgotten, as well as of the extent to which Western Europe and North America are lagging behind in the development of this therapy.

Knowledge and expertise are at the heart of the tensions in the differential histories of phage use, and it is in this context that the actions of those who travel to Georgia or Russia for treatment or to obtain phages should be situated. On one hand, it is a recognized therapy in the countries of the former Eastern bloc that has been used without interruption since its adoption in the 1920s; bacteriophage viruses are available in pharmacies, and there are no longer any doubts about their efficacy. Phages are medicinal products and exist as such, period. On the other hand, in the West, it is a therapy that has been relegated to the dusty shelves of research institutes and removed from textbooks and the *Vidal*, the "doctors' bible." It is a therapy that has continued to exist only discreetly, largely due to the provision of phages from the former USSR, and the situation may have continued in this way were it not for the increase in bacterial resistance to antibiotics that has made it necessary to seek alternatives.

COMING OUT

[The year] 2007 was supposed to be the "coming-out" year. Alain Dublanchet was active on all fronts because he had been looking for other sources of support for years, and the États généraux de la sécurité du patient et des infections nosocomiales [French National Convention on Patient Safety and Hospital-Acquired Infections] had become a major public event at which the press covers these subjects. In 2007, I therefore wanted to put the spotlight on phages. And then my scientific council threatened to resign—the very people who are now defending phages![18]

Alain-Michel Ceretti, the president of France Assos Santé who had shown sympathy toward André's anger and arguments at the March 2019 Temporary Specialized Scientific Committee, also founded Le Lien ("The Link"), an association devoted to combating, studying, and communicating about hospital-acquired infections. He created Le Lien to help victims of medical accidents in 1998, after the revelation of serious breaches by surgeons at the Clinique du sport in Paris that led to the infection of fifty-seven patients who had undergone lumbar or cervical surgery between January 1988 and May 1993. Ceretti became interested in phages in the first years of the twenty-first century as part of his commitment to combating hospital-acquired infections, which are often highly resistant to antibiotics and difficult, if not impossible, to treat. He then met Alain Dublanchet, a leading figure in phage therapy in France, who had spent several years striving to have this practice recognized by his colleagues in virology and infectious diseases. The two men spent several months analyzing the situation and developing arguments. The 2007 national convention for Le Lien seemed to provide an opportunity to openly promote bacteriophages for therapeutic purposes to not only their colleagues but also the general public. However, the scientists' efforts were thwarted by the members of the convention's scientific committee, most of whom were physicians. Dublanchet described this categorical refusal to me in an interview conducted in December 2016 in which he made no secret of his exasperation with the "crass ignorance" of his colleagues, whom he believed saw viruses only in terms of what they had learned from their training: as dangerous and deadly entities. He explained that many of his colleagues refused to recognize phage therapy because doing so would mean admitting their ignorance and calling into question their decades of experience in infectious diseases. Though harsh, Dublanchet's criticism reflects how viruses were considered by many health care professionals at the time—and still are to some extent, notably because of the COVID-19 pandemic.

Since this first abortive attempt, however, the situation has begun to change, as recounted by the founder of Le Lien. Former detractors are becoming advocates of phages, sometimes fervent ones. Scientific and clinical knowledge are being developed in France and abroad; people are being treated on a compassionate basis; and publications are discussing the increase in antimicrobial resistance, thus raising awareness of the pressing need for alternatives to antibiotics. In 2009, Ceretti and Dublanchet met with the CEO of a new start-up company called Pherecydes Pharma, which was entering the phage therapy market with the aim of developing and marketing bacteriophage viruses for use in humans. We seemed to be witnessing a genuine revival of phage therapy in France. Ceretti, who was a member of the board of a major medical foundation, even tried to convince a prominent pharmaceutical company to visit the start-up and invest in phage therapy. But things did not go as planned. He told me, "They're not going to follow this path because nothing is patentable. And that's where I started to understand the problem. In fact, the key issue previously was not patentability; it was use, importation, or manufacturing. In the meantime, it has become a medicinal product. Manufacturing it becomes complicated, and importing it from a country where there are no clinical studies is impossible. . . . I see how complicated it is, and so does Alain."

After disappearing from the therapeutic arsenal in the West at the end of the 1970s, bacteriophage viruses reappeared in European and American health product regulations in 2011. Indeed, the need to grant them the status of a medicinal product emerged because of the scientific and technical advances being made, the increase in bacterial resistance to antibiotics, and the subsequent increase in demand from scientists, doctors, and people with antibiotic-resistant infections in Western countries. Phages are therefore now considered medicinal products in the United States and the European Union. The European Union defines a medicinal product as "a substance or combination of substances that is intended to treat, prevent or diagnose a disease, or to restore, correct or modify physiological functions by exerting a pharmacological, immunological or metabolic action."[19]

The status of bacteriophage viruses as medicinal products is accompanied by new requirements, to which any person or group wishing to use them in humans must now conform. First, phages must be produced in accordance with good manufacturing practices (GMP). GMP form part of the quality assurance system put in place by the pharmaceutical industry to ensure the production of drugs in a systematic, standardized, and controlled manner according to the quality standards relevant to their use. They are a set of strict standards and procedures designed to ensure the quality of the finished product. Only pharmaceutical companies can supply GMP-qualified products. Second, the safety and efficacy of the bacteriophages produced must be demonstrated in randomized clinical trials (which I will discuss in chapter 6). Finally, tested products must obtain a marketing authorization. In France, a marketing authorization is issued by the National Agency for the Safety of Medicines and Health Products (ANSM) after an evaluation of the application submitted by the pharmaceutical company. This documentation must include the results of clinical trials demonstrating the efficacy and the benefit–risk ratio of the entity and describe the manufacturing processes used. No authorization has yet been granted for the marketing of bacteriophage viruses in France. However, once a candidate drug is in the evaluation phase of a clinical trial, and has therefore been qualified under GMP standards, it is eligible for "early access authorization," which is reimbursed at 100 percent by the French health insurance scheme.[20]

In the absence of phage-based preparations that have been granted a marketing authorization or, at the very least, of "GMP-compliant phages," giving sick people access to experimental treatments is a possibility but incurs the ethical, civil, and criminal liability of the prescribing physician and the administering pharmacist.[21] This is what is referred to as "compassionate care" and is how most people treated with phages in France today receive their care. Although the ANSM is involved in these treatments and supports health care professionals with a view to supporting the development of phage therapy, the fact remains that care providers bear full legal liability despite the regulatory agency's knowledge of the origin (whether France or elsewhere in the European Union) and manufacturing method of the products used.[22]

These regulatory and technical constraints concern European Union member states, as well as the United States and Japan, all members of the International Council for Harmonisation of Technical Requirements for Pharmaceuticals for Human Use. This council, created in 1990, aims to establish the conditions for a global market by harmonizing regulatory requirements via the application of the principles of reciprocity and mutual recognition by its three reference authorities. As Russia and most Eastern European countries are not members of the council, health products manufactured in these countries cannot legally be used in the European Union without a number of preliminary tests and an inspection of production facilities by the regulatory authorities. The case of Poland is enlightening in this respect and regarding the materialization of the role of borders. Phage therapy had been practiced in Poland for decades, notably at the Hirszfeld Institute for Immunology and Experimental Therapy.

In 2005, after Poland joined the European Union, the institute was authorized to open a five-room clinic to treat patients with antibiotic-resistant infections. However, to comply with GMP, it had to contract with a local vaccine company to produce phages under strictly controlled conditions. Nonetheless, phages will continue to be considered experimental therapies as long as the quality of the evidence of safety and efficacy is considered insufficient, despite the experience gained by the Polish team in treating a number of diseases.[23]

Paradoxically, the granting of medicinal product status to bacteriophage viruses has "complicated" matters, as Alain-Michel Ceretti puts it, for as long as phages were not considered medicinal products, they could be imported and administered discreetly, with responsibility for the treatment residing with the medical staff caring for the sick person. The ontological change that accompanied their legal categorization as "medicinal products" entails compliance with previously ignored standards and rules and disqualifies phage preparations that do not meet them. As Ceretti explains, "In the meantime, it has *become* a medicinal product. Manufacturing it becomes complicated, and importing it from a country where there are no clinical studies is impossible." Phages available in Georgia and Russia can no longer be used given the lack of information about their manufacture. As for doctors wishing to supervise the administration of phages from Russia in their patients, the ANSM's sole option is to advise them against it and reminding them of the risks involved. However, phages that meet the requisite quality and safety criteria are still not being produced in France or elsewhere in the European Union. André's anger; the trips to Georgia; Alain Dublanchet's impatience, fueled by his fear of never seeing the advent of this therapy in France; and Alain-Michel Ceretti's weariness are engendered by these circumstances, circumstances in which their hopes and aspirations also reside.

Bacteriophage viruses have been considered medicinal products, were used in many countries for several decades, and were registered in the *Vidal* pharmaceutical dictionary until the late 1970s. However, there is nothing straightforward about their reintegration because what these (all-too) brief alternative histories tell us is that, "in the meantime," everything has changed, from knowledge of these viruses, how they are used, the types of infections people experience, and conceptions of disease to health product regulations, harmonization across countries, the structuring of the drug market, and the appearance of "super-bugs."[24] All of these factors, and many others, contribute to the distinction—the very incommensurability—between what therapeutic phages (if the same term can be used to cover extremely diverse entities and practices) were until the 1970s and what they are or could be today. Therefore, merely releasing them from their dusty vials is insufficient. We must reinvent them (figure 2.1).

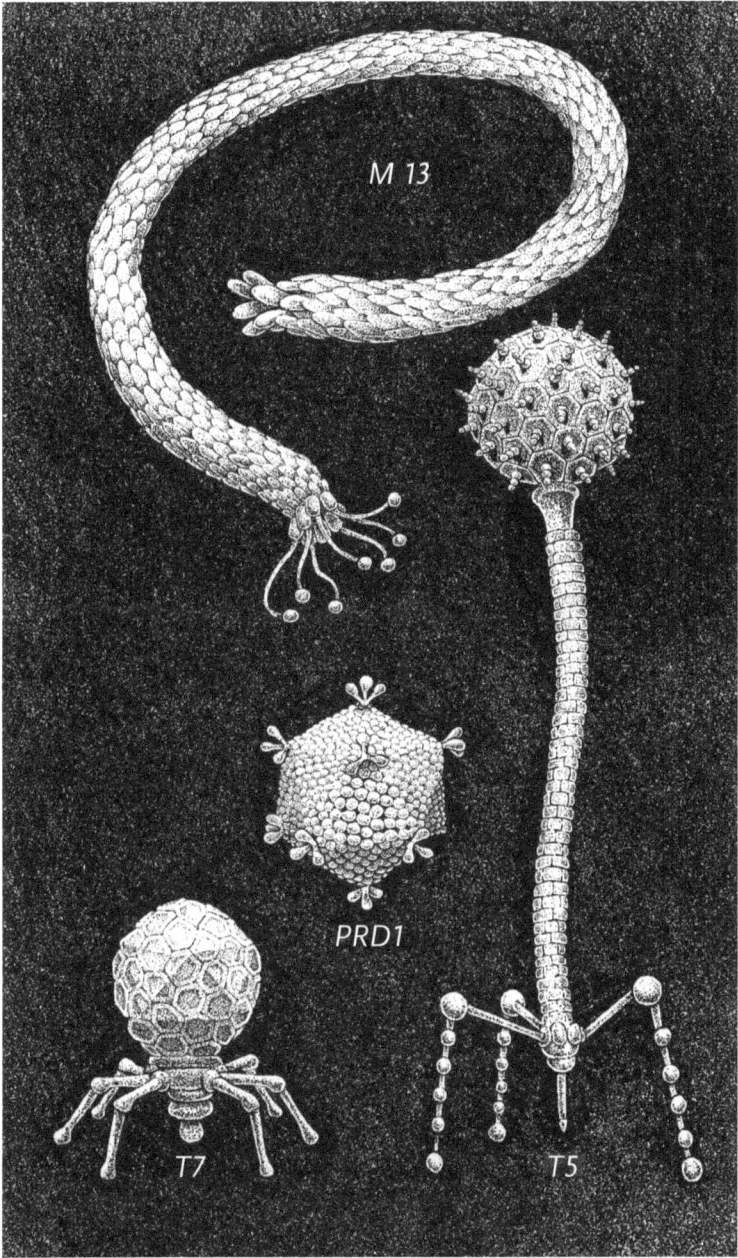

FIGURE 2.1 Several families of bacteriophage viruses

3

MICROGEOHISTORIES

Microbes evolve. They adapt. This is evidenced by the rise of antimicrobial resistance, as well as by the testimonies of people like André. Microbes adapt in environments; they adapt down to the very depths of the body's intimacy. Their nucleic acids—DNA or RNA—are marked by these adjustments, all corresponding to changes in their constituent series of A, C, G, and T nucleobases, and all representing traces—inscribed in their genetic codes—of the history of these microorganisms. Traces, also, of the human societies they pervade and in which they can live. This is what the now-famous COVID-19 variants tell us. This is also what the natural history of HIV tells us.[1] The properties of viral strains and the available sociopolitical data have enabled the reconstitution of the various trajectories of viruses and humans worldwide, pointing to the existence of many "situated biologies," as well as to singular multispecies or sociopolitical arrangements.[2]

Using phage therapy means taking advantage of the adaptive and coevolutionary capabilities of phages and bacteria. More precisely, it means using certain types of phages—virulent phages—capable only of performing lytic cycles (see figure 0.2 bottom), to treat bacterial infections. However, phages cannot interact with just any kind of bacteria. The two entities must recognize each other during the adsorption phase: A given phage can generally interact with only one bacterial species and typically with only certain genetic variants of that species. This great specificity requires access to a sufficiently high diversity of phages to ensure that at least one will be active on the bacteria to be treated and thus be capable of effectively treating a bacterial infection. Thus, before even considering treating patients, phage collections must be built up. However, this is a paradoxical task since it is based on capturing these incessantly evolving viruses in environments to immobilize them using procedures that must guarantee their immutability. How do scientists work with phages in the lab? Do they consider their evolutive capacities? If so, how?

Although bacteriophage viruses suffer from a lack of exposure to the general public and even to the life sciences, their capabilities and relational capacities have been used in laboratories around the world for decades, sometimes without the knowledge of the people using them. In 1950, Esther Lederberg discovered the *lambda* phage, which had properties previously unknown in other viruses, namely the ability to incorporate its genetic material into a bacterium's genetic material. Despite being fundamental to the discovery of the lysogenic cycle and related breakthroughs, Esther Lederberg's work sadly remains mostly ignored.[3] The "phage group," an informal network established at the end of the 1930s and led by Max Delbrück, Salvador Luria, and Alfred Hershey, among others, was responsible for many studies that laid the foundations of molecular biology. In 1952, Alfred Hershey and Martha Chase demonstrated that DNA was the vehicle for heredity by conducting an experiment on interactions between the T2 phage and the *Escherichia coli* bacterium. In 1965, the trio of André

Lwoff, François Jacob, and Jacques Monod won the Nobel Prize in Physiology or Medicine for their work on the "operon model" and their description of the structure and regulation of gene expression. This research focused on a phage. The first double-stranded DNA genome to be sequenced belonged to a phage. Some of the cloning techniques available in the laboratory are based on the use of the acquired knowledge of phages and their capabilities and on the use of their enzymes. Most recently, the awarding of the 2020 Nobel Prize in Chemistry to Emmanuelle Charpentier and Jennifer Doudna for their work on the CRISPR-Cas9 system, better known as the "molecular scissors," has further confirmed the importance of studies on these viruses. CRISPR-Cas is a system of recognition between phages and bacteria carried by nucleic acids. Scientists have learned to repurpose it to manipulate DNA by inserting or removing genes, thereby providing a simpler and safer way to perform what is referred to as "gene editing." The knowledge of phages and bacteria produced in this way has fueled the development of molecular biology and opened up vast research horizons, leading to applications that affect the daily lives of billions of people (e.g., genetically modified organisms, therapeutic molecules, biofuels).[4]

The laboratory practices described in the following pages are considerably simpler and much less spectacular. They are nonetheless fundamental for two reasons. First, they form the very basis of all phage therapy: Treatment with phages requires the availability of characterized, isolated, and purified viruses.[5] Second, focusing on actions, attitudes, and behaviors highlights the constraints of working with microorganisms, as well as the rules that must be followed and expertise required to work with them, the specificities of which must not be forgotten when the phages leave the laboratory.

BUILDING UP COLLECTIONS AND
CHARACTERIZING RELATIONSHIPS

Grégory Resch has been studying bacteriophage viruses for more than twenty years. When I visited his laboratory in the Department of Fundamental Microbiology at the University of Lausanne, his research was focused on their use in human therapy.[6] Although approximately two hundred bacterial species are pathogenic to humans, Grégory has chosen to focus his research on just five—but five that are particularly problematic because they belong to the category of what are now commonly referred to as "superbugs." These are bacteria that are multi-resistant or even completely resistant to available antibiotics: *Staphylococcus aureus, Pseudomonas aeruginosa, Acinetobacter baumannii, Escherichia coli,* and *Klebsiella pneumoniae.*[7] Because of this status, the World Health Organization considers them major health scourges.

For each of these bacteria, Grégory has collected several hundred strains, that is, genetic variants of the same species, which are characterized, sequenced, and then stored in small Eppendorf tubes in a freezer at –80°C. These bacteria mostly originate from samples taken from hospitalized patients. Accordingly, about two hundred strains present in the laboratory's *Pseudomonas aeruginosa* collection originate from samples taken from patients with cystic fibrosis hospitalized for chronic pulmonary infections. The germs responsible for such infections typically become resistant to antibiotics because of the numerous treatments administered to these patients since childhood. As a collection, these strains constitute the archive of infections of cystic fibrosis patients in a hospital with which Grégory's team regularly collaborates; each strain tells a story and becomes a record of a particular moment in a person's trajectory. These bacterial collections are doubly precious because of the often painful stories they illustrate and because their abundance and diversity make them unparalleled vehicles for studying the potentialities of phages.

The laboratory does indeed house phage collections or, more precisely, one phage collection for each bacterial collection, with samples of each phage also being stored in small Eppendorf tubes and kept in a refrigerator at 4°C and in a freezer at -80°C. Many of the 130 anti–*Pseudomonas aeruginosa* phages that make up the collection were isolated by Grégory's team (I will return to phage isolation later), and a few others were provided by several international research teams. An in-depth study of this collection would reveal the history behind Grégory's team: the changes in its composition, its research themes, and its collaborations with other scientists during various periods.[8] Consequently, if the bacterial collection of *Pseudomonas aeruginosa* constitutes a living archive of the hospital with which Grégory collaborates, then the phage collection constitutes an archive of the experimental history of Grégory's team.

When I visited Grégory's laboratory, most of the activity involved crossing these collections: bringing each bacterium in the "*Pseudo*" collection into contact with each phage in the "anti-*Pseudo*" collection to see what occurred during each encounter. This work, as well as that involved in the isolation of new phages, is essential because it enables the identification of the activity or host spectrum of each phage, that is, its ability to recognize and lyse various bacterial strains. In so doing, this work also promotes the development of new knowledge of and expertise on the relationships and dynamics between phages and bacteria, which are key to the development of a therapy based precisely on the relational capacities of these entities.

I spent most of my time with Julie, the team's technician who is responsible for almost all of the bench work, observing her manipulations of phages and bacteria and asking her questions about her every move.[9] Although, as she explained to me, many techniques are used when working with phages and bacteria, those employed when I visited the laboratory were relatively simple. Julie works mainly in solid media, that is, with Petri dishes. She first pours culture medium—a nutrient medium adapted to the microorganism she will be working with—into large square dishes, to which she has previously added one bacterium from the two hundred in the *Pseudomonas aeruginosa* collection. In a few minutes, the medium hardens to form a transparent agar. Normally, if the Petri dish is then placed in an incubator at 37°C—the optimal temperature for bacterial growth—the bacteria will reproduce to form a bacterial mat within a few hours. The agar will then take on an opaque appearance, proof of the presence of bacteria evenly distributed across the dish. If, before placing the dish in the incubator, drops of a suspension containing phages are placed on the hardened agar, "lysis plaques," or plaque-forming units (PFUs)—small transparent plaques, corresponding to holes in the bacterial mat—may be observed, indicating that there are no bacteria, or no longer any bacteria, in these locations. They have been lysed. Each lysis plaque is generally considered to result from the action of a phage that has attacked a bacterium, which then releases several hundred phages by lysing. These phages can then attack the bacteria in the immediate vicinity. This relatively simple experiment enables the measurement of phage activity: By counting the lysis plaques present on the dish, and by reducing this number to the amount of phage suspension deposited, one can establish the titer of the starting suspension, that is, the number of active phages that it contains, which is measured by number of PFUs per milliliter. In summary, the more holes (PFUs) there are, the greater the phage's capacity to induce lytic cycles with the bacteria and the more appealing it becomes from a therapeutic perspective.

Over the following weeks and months, Julie puts each bacterium in contact with each phage, corresponding to around twenty-six thousand crossings.[10] The day after the phages are deposited on the Petri dishes, Julie "reads" the results. Sitting in front of her computer, surrounded by Petri dishes prepared the day before, she first takes photographs of all the dishes, which she then transfers to the computer. Next, she observes each one in detail. Facing a window, she removes the cover of a dish, grasps it firmly between the thumb and forefinger of her left hand, places it at eye level, and, with her right hand, makes a mark on each lysis area, which she observes with a fine-tipped marker while counting. She also makes annotations, which were totally incomprehensible to me at first. For several dozen minutes, her brow furrowed in concentration, Julie scrutinizes the dishes. For some, this operation takes only a few seconds. For others, she hesitates and changes the dish's orientation. She counts and, with speed acquired over long hours spent working with phages and bacteria, almost immediately notes the titer (the number of phages present) in the initial solution. Her pen glides from one row to another, from one column to another, sometimes drawing symbols and adding numbers. Although I can intuitively understand which phages are active and which are inactive or barely active simply by looking at the dishes, her gestures and attitudes seem to reveal something much more subtle.

The numbers correspond to the quantification of the number of phages in the starting solution, but other inscriptions are also added: "OL" and "SCL." "OL" stands for "opaque lysis." In this case, the lysis plaques are visible but slightly opaque. More precisely, they are less opaque than the bacterial mat but not transparent. Transparent lysis plaques are referred to as "clear." Between the two comes "SCL," meaning " semiclear lysis." Julie then goes further, noting the presence of bacterial regrowth on the lysis areas: On lysis plaques sometimes measuring less than a millimeter across, her expert eye is capable of discerning the presence of small opaque spots signifying the presence of bacteria that have multiplied again. Here, some of the bacteria have not been lysed by the phage and have therefore developed a resistance to it, allowing the bacteria to reproduce again, even in the presence of the phage. This information is particularly important for the therapeutic use of phages: These viruses can now be excluded. However, Julie can also identify what she calls "double morphologies," that is, lysis plaques whose appearance differs. Her trained eye can tell the difference between lysis plaques that are not exactly the same shape or size. This difference in appearance implies the presence of different phages and therefore gives an indication of the "purity" of the phage solution used. If this solution contains a majority of a single phage, the presence of lysis plaques with different morphologies implies that a minority proportion of another phage is also present, and provides a reminder of the ever-present uncertainty in a microbiologist's life: Unless we actively seek out the presence of an invisible entity, nothing can ever guarantee that an occasional undesirable bug has not gatecrashed the party.

Julie's annotations characterize not only a quantification of a phage's activity but also a *quality of interaction*. Her inscriptions objectify a relationship between a phage and a bacterium. Such an objectification is all the more surprising since, in this experiment, *the phage is not visible at any time*. It is the bacterium that testifies to the presence or absence of a phage and its activity, and it is also the bacterium that testifies to the quality of the relationship. The phage's presence is evidenced by the relative absence of the bacterium and the form that this absence takes on the agar of the Petri dish.

Julie then enters the results of these crossings in a seemingly simple table with one column for each phage, one row for each bacterium, and a box that summarizes their relationships, enabling—via color coding and the use of different fonts—a *quantitative and qualitative* assessment of the relationships between phages and bacteria at a single glance. Once the data has been entered into the table, the dishes are disposed of. Only the stock solutions of each phage and each bacterium are retained, stored individually in refrigerators at 4°C and in freezers at –80°C. The photographs and data table are therefore the only records of these interactions. The entire chain of ephemeral inscriptions specific to reference work in the laboratory disappears in favor of the table, which objectifies the relationships between phages and bacteria.[11] In this series of experiments, we are witnessing a *trend toward the objectification* of interspecies relationships.

The photographs are preserved carefully. They are important reminders that what happens in the laboratory is more than an interaction between phages and bacteria and that the objectification of their relationship involves a third term: the technician. The photographs are used as a supporting resource by Grégory and Julie, who examine them when they consult the table. When I put the hypothesis to Julie that the observations are homogeneous because they are all made by the same person, she responds, "Not necessarily. In the vast majority of cases, yes, but the brightness of the light plays an important role. Eye strain is also a factor. Here, you see, there's not too much light outside, so I will see the results in a certain way, but if the sun comes out, in the time taken to enter the results, my results may vary just a little." This point is widely emphasized in laboratory studies, as well as in studies involving living entities in general. Objectification is a process that requires adjustments and negotiations, one that depends partly on the expertise that humans have developed in their interactions with microorganisms and on the broader context in which the objectification takes place.[12] This is not a question of relativism; it is simply a matter of reiterating the processual, relational, and contingent dimension of knowledge production. It is these adjustments, and Julie's intimate knowledge of phages and bacteria, that will enable the creation of a matrix that will finally objectify some twenty-six thousand relationships: all different, all specific, and all situated.

5055-KP98, THE ISOLATION OF NEW PHAGES, AND ENTRY INTO THE COLLECTION

Shortly before my arrival at the laboratory, Grégory received two strains of bacterial species: one strain of *Klebsiella pneumoniae* (KP) and one strain of *Proteus mirabilis* (PM), both resistant to antibiotics and both responsible for the infection of a person being treated in a French university hospital. After the failure of successive treatments, the doctors in charge of this person's care turned to Grégory with a view to attempting a phage treatment. However, the laboratory lacked phages capable of counteracting *Proteus*, a bacterium they do not normally work with, and the phages they possessed for *Klebsiella* proved ineffective for the patient's strain. Grégory and Julie therefore decided to find phages active against these two bacteria so that they could be sent to the university hospital and administered to the patient.

To this end, Julie carried out tests with various water samples, which she referred to as "passages d'eau." As in the first set of experiments, she poured a nutrient medium into several Petri dishes and added her bacterium of interest: either KP or PM. She then poured water from various sources onto each dish. As mentioned, phages are strict parasites of bacteria and are found wherever bacteria live. The best place to find them, however, is in "rich" water, such as sewage or water from treatment plants: The more bacteria there are in a milieu, the more opportunities there are to find phages. Over the past few years, however, I have learned that not all water is considered equal. During discussions with a technician working for a start-up company, I learned that samples should not be taken downstream from chemical plants, which often discharge toxic substances and biocides into waterways. This practice complicates efforts to find any traces of life, even microbial. Julie favors hospital wastewater. Since hospitals are hotbeds of bacteria, most of which are multiresistant owing to the strong selection pressure exerted by repeated antibiotic treatments administered to patients, there is a high probability of finding active phages. One researcher confided to me that she preferred water from sewage treatment plants or from ponds and puddles to that of hospitals, considering that the high presence of antibiotics in hospital wastewater signifies a lower presence of bacteria and therefore of phages. Water treatment techniques and hygiene habits may also play a role in the amount of bacteria and phages found in wastewater.

In any case, deciding on phage collection sites above all implies defining the milieus most likely to harbor an abundant population of bacteria. However, as was highlighted in the crossing of collections, the extremely high specificity of phages, stemming from their long coevolutionary relationships with bacteria, also leads some of the people I met, including Julie, to consider the *territorialization* of collection. These experts believe that it is most fruitful to choose collection sites that are as close as possible to the geographical origin of the bacteria of interest. As Julie put it: "Most of the time this will be easier because . . . there will be an entire surrounding population with potentially the same bacteria. We realize that, in general, our bacteria in Europe will not be the same as in the United States, so the water samples will not function in the same way. You are unlikely to get a phage from the United States to actually work on your own collections here in Europe."

To illustrate the territorialization of relationships between phages and bacteria, Julie describes the case of a patient hospitalized at the Hospices Civils de Lyon, Lyon being a city relatively close to Grégory's laboratory. The Lyon team asked several European laboratories (in France, England, Spain, Portugal, and Belgium) to find a phage active against the bacteria responsible for this patient's infection. Only Grégory's laboratory had phages that showed satisfactory activity on the patient's bacteria, all of which had been isolated in local water.[13] This example, which highlights the ecosystemic dimension of phage therapy, is often mentioned by researchers I have met in recent years. Although the entire community is not as adamant on this point as Julie, with some scientists having successfully experimented with interactions between phages and bacteria from geographically distant areas, everyone acknowledges the importance of territorialization resulting from the coevolutionary interactions between phages and bacteria.

Julie bears all these hypotheses in mind as she deposits this bacteria-rich water on her Petri dishes, or rather, it is by dint of placing bacteria-rich water on dishes, and repeated observations, that Julie develops explanations and arguments concerning the "richness" of water samples, and the links between micro and macro, between the biology of organisms and human communities. Once the water has been deposited on each Petri dish, Julie places them in an incubator at 37°C for twenty-four hours. The next day, she carefully examines the dishes. Wherever she sees a lysis plaque (signaling the possible presence of a phage), she takes a sample. She next prepares a screw-cap tube in which she puts a little liquid nutrient medium, the bacterium of interest, and the sample taken from the lysis plaque. She then places it in the incubator for three hours. After this time, during which the bacteria (and perhaps the phages) have reproduced, she centrifuges the tube for ten minutes. The heavy bacteria are concentrated in a small deposit at the bottom of the tube known as the "pellet." Julie recovers the supernatant, which should contain the phages, and filters it to eliminate the bacterial debris. Finally, she deposits some of this filtrate (and again, perhaps, some phages) onto a Petri dish containing nutrient medium and the bacterium of interest and places the dish in an incubator at 37°C for twenty-four hours. If any lysis plaques are still apparent, she will start the entire process again: sampling, incubation, centrifugation, and filtration. She may make three to eight passes, depending on what she observes. Julie tells me that these successive passages d'eau allow her to verify that she is dealing with a single type of phage, rather than two or three. As we learned in the previous experiment, a single lysis plaque may well be produced by the action of different phages.

During these various cycles, Julie learns a lot about her phage and her bacteria: the best moment to put them together, the time required for interaction, and the optimal temperature. Although the methods are standardized and the techniques seem simple, what she has learned to consider are not the characteristics of a phage and bacterium but rather the characteristics of their encounter. This is another reminder that each encounter and each relationship is unique: "You're proceeding by trial and error! You have run tests; you can try altering the temperature a little if you can. It's all about trial and error, you know. So, all right, everybody says it's really easy [to produce phages], and on the whole, yes, it is easy, but afterward, you may need to make adjustments. And it's true that wanting to produce all your phages in a particular way [is difficult]. . . . You proceed on a phage-by-phage basis." "On a phage-by-phage basis." In five years of fieldwork, I recorded no other occurrence of this phrase, yet it sums up the practices and discourses related to phage therapy succinctly. Intermingled with objectified knowledge and noted in summary sheets, laboratory notebooks, and further downstream in scientific papers are the knowledge and know-how that Julie has developed through her contact with microorganisms and her highly detailed knowledge of the phages and bacteria in *laboratory collections*, thus making Julie's acquired expertise indispensable. Some bacteria, such as *Staphylococcus aureus*, are described as "capricious," whereas others are described as "permissive." Some do not "work properly," which can lead to "your phage production being less effective." It is therefore a question of taking the particularities of each relationship into account, particularities linked to the manner in which history is embodied in the biology of each organism.

After the passages d'eau, Julie obtains three phages active against the patient's strain of KP and two phages active against their strain of PM. Although she will sequence these phages to determine their genomes, she is already convinced that the three anti-*Klebsiella* phages are different because, as she points out, in addition to having been isolated from water of different origins, "they have a completely different morphology. You see, typically, I think that for KP98, I really have three different phages. Those ones are good. This one [one of the three phages] is very tiny; it has a touch of OL [opaque lysis]; as you can see, it's not very clear." This formulation—"they have a completely different morphology"—is in fact a short cut because it is actually the lysis plaques, and therefore the holes in the bacterial mat— rather than the phages themselves—that do not have the same morphology. The presence of phages in the lab is most often assumed, as when Julie tries to isolate new ones: They *may be* in the water deposited onto the Petri dishes; they *may be* in a lysis plaque; they *may be* in the process of lysing bacteria in a tube; they *may be* in the supernatant. It is the repetition of the passages d'eau and Julie's keen eye for evaluating small holes in the agar or the turbidity of solutions in tubes that provides clues to the presence of these entities, which are always hidden from view.

What is evaluated is not a biological entity as such but rather a *relationship* between viruses and bacteria. The presence and capacities of the phages are confirmed only by the disappearance of bacteria, and this confirmation is highly dependent on the technician's interpretations. In the following weeks, Julie will further characterize the relationships among the phage, the bacteria, and the environment in which they interact.[14] In particular, she will determine the lysis time, that is, the time between the initial contact between the phage and the bacterium and the appearance of the first bacterial lyses, as well as the "burst size," that is, the number of new viral particles—new phages produced by each bacterium—all of which are valid only for the specific phage–bacteria pair in question.

This existence by relationship is illustrated by the attribution of a name to each newly isolated phage; for example, one of the anti-*Klebsiella* phages is named 5055-KP98. All phages in the lab with names beginning with "5000" are KP phages; 5055-KP98 is thus the fifty-fifth KP phage in the collection. The "KP98" corresponds to the name of the bacterial strain from which it was isolated. Since phages and bacteria are constantly coevolving, if this phage were to come into contact with another bacterium—KP112, for example—the resulting phage would be given the name 5055-KP112 to reflect its probable evolution. It would definitely no longer be the same phage. As Julie explains, "If you change your bacteria, or anything else, your results will be different. Your phage is not going to react in the same way; it's a phage–bacteria pair."

Therefore, the attribution of a name does more than mark the scientific "birth" of a phage. The name provides a snapshot, at a time t, of a relationship that is at least tripartite and destined to evolve permanently. It is a record of what I have chosen to call *microgeohistories*: a bacterial strain whose capacities (e.g., resistance, virulence) have been acquired and developed partly during its encounter with a human body, a fragment of a person's biography, the fruit of a trajectory and multiple relationships at a given time and place; a phage collected from the wastewater collection system of a hospital's infectious diseases department, in which it forged relationships with one or more bacteria that contributed to making it what it was at the time of the encounter with this strain; two entities brought together by a technician who has learned from her experience of creating interactions between other microorganisms, long hours spent observing Petri dishes and Eppendorf tubes and formulating hypotheses about optimal temperature and contact time.[15] Phage 5055-KP98, which has joined the collection of the Department of Fundamental Microbiology at the University of Lausanne, bears traces of the times and places encountered by phages from the previous cycles from which it originated and the fruit of its encounter with the KP98 strain in a medium been partly determined by the technician—all inscribed in a succession of nucleic acids embedded in its genome. Choosing the name 5055-KP98 is above all an act of "freezing," of crystallizing and essentializing a relationship, situated in time and space, within biological entities.

RECALCITRANCE

However—and this is precisely what is at stake—this colossal task of isolation and characterization must not have been in vain: Once the phage has been incorporated into the collection, there can be no further changes. While remaining alive, the bacteria from patients and the phages from wastewater must not change or evolve, neither in test tubes nor on paper, because the knowledge acquired is valid only for this specific pairing, at this specific moment of their relationship, in the specific environment in which the relationship occurred.

However, such objectification is never fully achieved. Research on laboratory organisms—whether fruit flies, human cells, viruses, yeast, mice, or plants—not only describes the modifications made to the organisms and the conditions of their existence but also bears witness to the organization and moral economy of the laboratory required to make these organisms "living technologies," to use a phrase coined by the historian of science Robert Kohler. These are entities that "do things that humans value but that they might not have done in nature."[16] The work of standardizing and normalizing biological entities is based on a division of scientific activity, on rules and implicit norms specific to each organism. This work forms the basis of all generalizable knowledge because it enables comparisons to be made across experiments conducted at different times or places.[17] If two laboratories, thousands of kilometers or two decades apart, are working on an "identical" organism, then the knowledge they will produce concerning this organism will be comparable. In this standardization and normalization of laboratory organisms, cryogenic preservation techniques have taken on paramount importance in recent decades: Freezing an organism, under the right conditions and in the right environment, makes it possible to momentarily suspend the movement inherent to all life and consequently to transport the organism over spatially and temporally vast distances from one laboratory to another.[18]

In a room separated from the manipulation rooms in which I shadowed Julie for a few days—a physical separation that substantiates an ontological difference between the preserved entities—there are two $-80°C$ freezers containing the lab's precious collections of phages and bacteria: polystyrene boxes containing Eppendorf cryotubes, each housing either a phage or bacterium, according to the freezer. This series of containers is connected to alarms configured to alert Grégory's team should the freezer fail to fulfill its intended role: keeping all these microorganisms at a temperature that prevents them from moving, deteriorating, and mutating; freezing—for an indefinite period—microgeohistories whose value is further increased by the fact that they form the basis of the development of a therapy that could reduce or even eliminate the suffering of people living with chronic infections, many of whom have been suffering for years or even decades.

In this story, however, there is a great danger of forgetting how the collections were actually produced and assembled, and of only considering phages as scientific objects that could easily become therapeutic entities. The presence of two refrigerators (one for phages, one for bacteria) to prevent cross-contamination; the rules for disinfecting hands before and after every experimental task; the various wastebaskets and their specific uses; the many annotations on tubes, benches, sheets, and notebooks; the skillful stacking of agar dishes in the laboratory incubator; the manner of drying the agar dishes under the hood; the perfectly aligned racks holding the tubes to simplify manipulations and prevent errors of identification; the gestures patiently acquired by dint of extensive experience—these are all precious indications of the types of rules put in place to enable staff to work with these phages and bacteria.[19] These rules are also subtle indications that what goes on in the laboratory, far from being a story of human mastery over microorganisms, is much more akin to a succession of adjustments and negotiations.[20]

This is the case because, although freezing enables the partial suspension of time and movement, that suspension is only momentary. It is important to understand that phages remain dynamic, evolving entities always on the lookout for different conditions and new encounters. Living beings evolve, mutate, and adapt their habits and behavior. They are in motion, varying in both their biological composition and their interactions. The biological entities used by laboratories, despite being so highly standardized and normalized, therefore manifest a certain *recalcitrance* toward the projects reserved for them by the humans responsible for their domestication.[21] They are likely to do something other than what is expected of them, given their rich potentialities. But the phages in collections are not meant to remain confined to laboratories. They are meant to be used *outside the laboratory*, in the bodies of sick people (and perhaps beyond, in the bodies of nonhuman animals or in crops, for example), far from the conditions under which they were isolated and characterized. Therefore, before examining the possible clinical uses of these collections, we must understand what phages can do when they are not in the relatively controlled environments of lab benches and Eppendorf tubes.

4

PLURIBIOSIS

*Despite their paramount importance to human health, to sci-
ence and to all life on the planet, the phage field remains a niche
area of study. One reason that phages (as well as most viruses
that do not make us or our domesticates sick) remain over-
looked is that you can't just go out or look inside and observe
them. When outside a host cell, they travel as virions so small
that seeing them requires an electron microscope or other sophisticated and costly equipment. Most can't be cultured and inter-
rogated in the lab because their hosts are not known or not yet
culturable. . . . Most scientists and others just don't think of them
as alive. So this major component of life is reduced to its inert
intercellular transport form that is then subjected to biochemical analysis and described in lifeless terms, leaving us blind to
their nature as active agents. This is somewhat of a travesty, as
these bits of biochemistry are the most successful predators on the
planet. They are promiscuous and engage in kinky sexual games
(e.g., homologous and illegitimate recombination with related
and completely alien genomes, orgies of hundreds of genomes).
Humans observing virions perceive them to be inert, but these
"inert" particles, given contact with a potential host, reveal their
true nature as complicated nanomachines primed for action.*

This quotation, taken from a collective work by scientists renowned for their research on bacteriophage viruses, differs from André's testimony but is just as instructive concerning the difficulties of grasping the particularities of these entities, this time from a life sciences perspective. In just a few lines, the technical, material, and conceptual difficulties of working with phages are described, in addition to the prerequisites, prejudices, and orientations that guide scientific developments, from bench work (working with "inert entities" differs from working with "active agents") to research policies (little research is carried out on viruses that do not make humans sick), along with the moral values that accompany researchers in their daily work (the "loose morals" of these entities, which freely mix their genomes with anyone or anything). These difficulties are highlighted by the profusion of often virile characterizations indulged in by the authors: "virions," "active agents," "bits of biochemistry," "predators," "'inert' particles," "complicated nanomachines primed for action."

There is nothing exceptional about this profusion of terms: It is an everyday occurrence in scientists' activities—testimony to the many potentialities of living beings—because each example calls for another word or expression to facilitate the understanding of behaviors, phenomena, or entities. In the field of phage therapy, bacteriophage viruses are often presented as "snipers" or "professional killers" of bacteria; however, as shown in chapter 3, delicate adjustments are required for encounters between phages and bacteria. Words have meaning, and their uses have consequences. In this chapter, I consider what we know today, across several disciplines, about these entities and about the importance of naming things and phenomena.

PHAGES ARE THE NEW BACTERIA

"Wherever you find bacteria, you find phages." Many scientific papers begin with a sentence like this. For example, "Bacterio-phages are found infecting bacteria present in the oceans, fresh-water or saltwater lakes, different compartments of vertebrates such as the intestine, oral cavity, the skin or lungs, but also in insects, especially those hosting symbiotic bacteria. Bacterio-phages are also found associated with bacteriomes (bacterial populations associated with a particular ecosystem and com-posed of multiple species) of plants, whether in the rhizosphere, the roots or the aerial parts."[1]

Although the fundamental role of bacteria in all living beings, and for the great (bio)geochemical cycles, is now well established, it took great perseverance for the microbiologist Lynn Margulis to make her voice heard. In the 1960s, Margulis developed endo-symbiotic theory, which postulates that eukaryotic cells (cells with a nucleus containing genetic material, such as the cells of the human body) are the result of symbiotic associations with prokaryotes (cells without a nucleus, such as bacteria).[2] Mito-chondria, the tiny energy-generating organelles found in profu-sion in the eukaryotic cells of animals, are in fact ancient bacteria or fragments of bacteria that were phagocyted by the ancestor of eukaryotic cells. A similar story seems to have occurred for chloroplasts: These organelles, which enable photosynthesis to take place in eukaryotic plant cells, are also ancient cyanobacte-ria that were integrated during the evolutionary process. These theories, now widely accepted, make bacteria key players in the evolution of living beings. The rapid development of metage-nomics in the early years of the twenty-first century, allowing for the sequencing of DNA contained in samples taken from envi-ronments, made it possible to document and hypothesize many

other roles of bacteria in the ecosystems in which they participate. Research on what is referred to as the "microbiota" (the set
of microorganisms living in a specific environment) is contributing to the momentum generated in this field. For example, an
important paper published by the biologists Scott Gilbert and
Jan Sapp and the philosopher Alfred Tauber has established that
humans are dependent on the incredible quantity and diversity
of microbes that inhabit them, from immunological, developmental, genetic and physiological perspectives.[3] A convinc-
ing illustration of this is provided by laboratory experiments
showing that axenic mice—mice deprived of their microbes—
develop numerous pathologies and soon die. The evocative titles
of the growing number of books written for the general public
over the past few years (e.g., *Tous entrelacés!*, *I Contain Multitudes*, *Jamais seul*) remind us of the complex, intertwined links
between living beings and microbes, and of the need for multispecies relationships.[4]

"Wherever you find bacteria, you find phages." This sentence
suggests an endlessly repeating story for it indicates that we now
have to contend with an additional entity; that everything we
know, or think we know, about bacteria must now be interpreted
in light of their relationships with the bacteriophage viruses that
accompany them. This is why, after our visit to the laboratory, I
now propose another detour, this time into the world of microbial ecology. This detour is necessary because the entities that we
encountered in the lab freezers—in that quiet room away from
the tumult of experiments, interactions, and lysis plaques—were
reified entities. However, it is very easy to forget how they were
isolated and frozen in ice since, for the biology community,
phages most commonly serve as simple tools rather than biological entities in their own right. As the biologist Merry Youle
points out in *Thinking Like a Phage*, "Intensive exploration of a

few phages has enabled researchers to develop tool sets for working with them and thereby to penetrate deeply into some basic
cellular activities. On the other hand, this practice has tended to
establish paradigms that delay the discovery and appreciation of
the full diversity of phage tactics."[5] Behind the simplicity of the
model lie not only the diversity of phages and of their capacities
and abilities but also the various modes of relationships between
phages and bacteria, implicitly referred to by Youle as "tactics."
The scientists with whom I have been in contact for several
years share this view.

Phage ecology as a discipline was born only in the early 1990s,
but the already abundant body of available data allows us to
assess the roles of these viruses in various environments. In 1999,
Steven W. Wilhelm and Curtis A. Suttle described a complex
mechanism they called the "viral shunt": Every second, approximately 10^{23} viral infections occur in the ocean, leading to the
deaths of many of the infected hosts, from bacteria to whales,
and thus influencing the composition of marine communities.[6]
Bacteriophage viruses are major actors in the mortality of marine
bacteria, transforming living organisms into dissolved organic
matter and ensuring the availability of their constituent nutrients (especially carbon and nitrogen). The viral shunt is therefore of paramount importance to marine ecosystems and, more
globally, to biogeochemical cycles, including the carbon cycle:

> By the simplest approximation, the *viral shunt* moves material
> from living organisms into the particulate and dissolved pools of
> organic matter, where much of it is converted to carbon dioxide by
> respiration and photodegradation. However, the effects can also
> be more profound and potentially include the release of dimethyl
> sulfide, a gas that affects the Earth's climate. . . . Ultimately, it is
> both the quantity and composition of the material that is released

by viral lysis that affects microbial communities and global geo-
chemical cycles. As well as increasing the amount of respiration
in the system, the shunting of organic material from organisms to
the dissolved pool by viral lysis potentially influences the amount
of carbon that is exported to the deep ocean by the "biological pump." This is a globally significant process that sequesters
approximately three gigatons of carbon per year.[7]

This mechanism is also studied in microbial soil ecology to gain
a more precise understanding of the fate of nutrients in terrestrial ecosystems.

However, it is not always easy to "visualize" and appropriate the potentialities of phages and the effects of their actions
on environments. After I had spent a few months working on
this topic, Claire Le Hénaff-Le Marrec, a microbiologist, col-
league, and friend who has frequently accompanied me in my
discovery of these viruses, introduced me to a paper that provides an insight into the macroscopic effects of these invisible
entities without recourse to an electron microscope. Every year,
Lake Nakuru in Kenya hosts huge colonies of lesser flamingos.
Hundreds of thousands of these birds come to feed on *Arthrospira platensis*, cyanobacteria that are particularly abundant in
the saline and alkaline waters of the lake. In 2009, however, the
flamingo population dropped sharply from more than a million individuals to just over a thousand between the months of
June and October. This unusual decline was quickly correlated
with the disappearance of *Arthrospira*, the flamingos' main food
source. Samples routinely collected from the lake then revealed
the presence of the largest number of viruses ever reported in
a natural aquatic environment, which decimated the cyanobacteria populations by reproducing within them. The flamingos,
deprived of food, then flew off to more favorable places. This

is the first documented example of the influence of viruses on a food chain.[8] It is a telling example but one that should not lead us to draw simplistic conclusions.

Until now, I have discussed only the lytic potential of phages and the role of that potential in ecosystems and in major biogeochemical cycles. I have focused only on how phages can contribute to the formation and transformation of their environments by destroying bacteria. However, even though phages are most commonly presented as "bacteria killers" because of their lytic potential—which is precisely why they have attracted the attention of physicians in the context of phage therapy—they are also involved in many other types of interactions with bacteria and with their environments. This is the case for temperate phages, which can integrate their genetic material into the DNA of their host bacteria in the form of prophages as part of a lysogenic cycle (see figure 0.2). In this case, the genes carried by prophages engage in complex interactions with the genes of the host bacteria, whose potentialities are then transformed. This is because the prophages give the bacteria that carry them abilities and competencies that they previously lacked. Increased bacterial pathogenicity is among the best-documented cases of these novel abilities (not because it is the most common case but simply because these capacities are of foremost concern to humans). In this way, the toxins of *Vibrio cholerae* (the bacteria that cause cholera), *Shigella* (the bacteria that cause shigella infection, or shigellosis), *Corynebacterium diphtheriae* (the bacteria that cause diphtheria), and *Clostridium botulinum* (the bacteria that cause botulism) are secreted by bacteria but genetically encoded by the prophages they carry.[9] The same applies to some of the genes that confer antibiotic resistance in bacteria (as I will discuss in chapter 8).

In addition, lysogeny may be only a transient state. In the event of environmental stress, such as a change in salinity, temperature, or the presence of specific chemical molecules such as pollutants or antibiotics, the lytic cycle may be induced. (As hypothesized by the authors of the flamingo paper, it is possible that a change in the temperature or salinity of Lake Nakuru or an increase in UV levels could have induced a lytic cycle in an *Arthrospira* prophage, causing a massive infection and lysis of this bacterial population and leading to the local disappearance of the flamingos.)

However, during this transition, which implies a radical change in mode of relationship, the excision of the phage genome from the bacterial genome is not always "perfect." Sometimes the phage will take a piece of the bacterial genome with it, or it will leave a piece of its own. And, as discussed in chapter 3, these two entities—phages and bacteria—are constantly coevolving. This coevolution can occur through one-off mutations or through the sometimes massive transfers of genetic material referred to as "horizontal gene transfers" (as opposed to vertical gene transfers, which take place through filiation toward the descendants). There are different kinds of horizontal transfer. *Transduction*, for example, is the passage of a gene from one bacterium to another via a bacteriophage virus (figure 4.1). *Transformation*, on the other hand, occurs when lysed bacteria release DNA fragments that are "ingested" by other bacteria located nearby (figure 4.2). No further detail is required here; it is sufficient to point out that these changes, these mixtures and assimilations, occur at a prodigious rate and are an incredible force driving the evolution of phages, bacteria, and, potentially, other species. This is a force that biologists are only just beginning to grasp: Despite the constant discovery of new genes thanks to the unprecedented development of genetic sequencing—particularly metagenomics— the vast majority of phage genomes remain unknown, as do the genes they carry and the functions associated with them.

Transduction

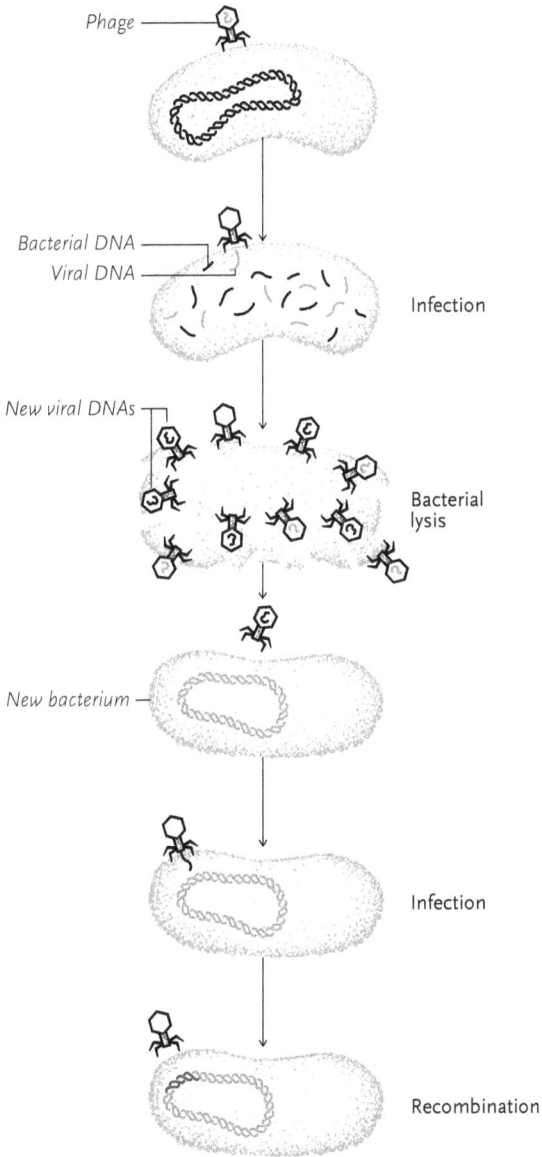

Phage

Bacterial DNA
Viral DNA

Infection

New viral DNAs

Bacterial
lysis

New bacterium

Infection

Recombination

FIGURE 4.1 Transduction

Transformation

Transformation by a plasmid

Bacterial chromosome
Plasmid

Plasmid
uptake

STABLE
TRANSFORMATION

Transformation by DNA fragments

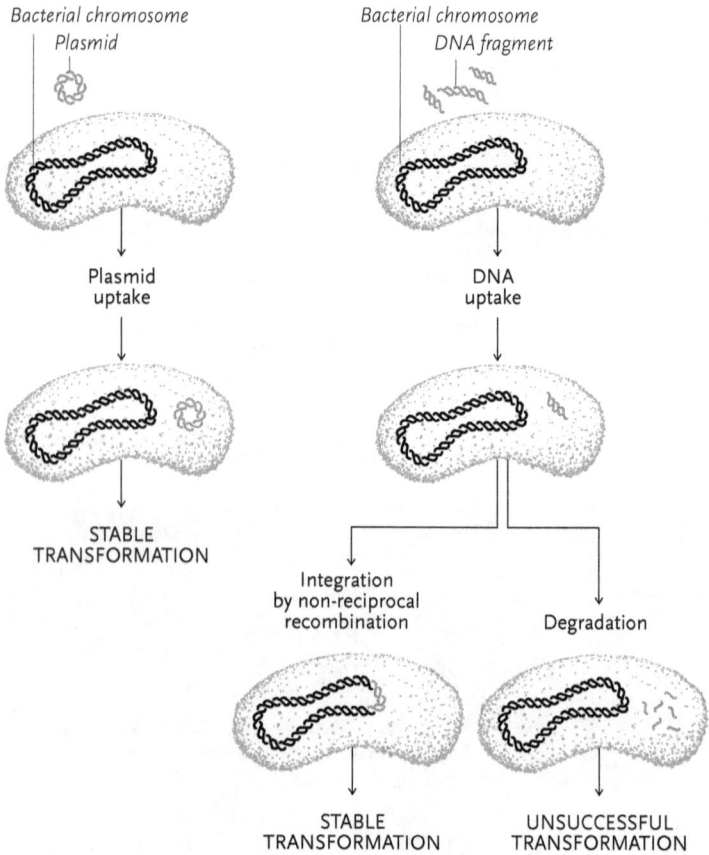

Bacterial chromosome
DNA fragment

DNA
uptake

Integration
by non-reciprocal
recombination

Degradation

STABLE
TRANSFORMATION

UNSUCCESSFUL
TRANSFORMATION

FIGURE 4.2 Transformation

Bacteriophage viruses thus form the largest pool of unknown genes on Earth and therefore the largest pool of unimaginable possibilities. Microbiologists thus refer to these unknown genes as "dark matter."[10]

VIRUSES AND METAPHORS

Bacteriophages, these composters of worlds and gene shifters—entities with phenomenal potentialities—are nonetheless frequently portrayed as "snipers" or "professional killers" that "inject" their DNA, "penetrate" host cells, and "hide" themselves, becoming "sleeping agents" in bacterial DNA. As for the mechanisms of coevolution between phages and bacteria, they are most often described as being involved in an "arms race."[11] The vocabulary used reflects transgression, violation, insidious conflict, and destruction, suggesting that battle and competition are the favored modes of interaction between phages and bacteria. However, phage behavior is also often described, sometimes by the same researchers, in very different terms, as when their potentialities are referred to as "incredible" or their sexual practices are called "kinky." These lexical uncertainties are also on display in the chapter-opening quotation from *Life in Our Phage World*. They are symptomatic of the difficulty of linguistically encapsulating what these entities are believed to be and do without being *reductive*. The authors of *Life in Our Phage World* are all the more aware of this because they know that the terms, categories, and conceptualizations used influence the production of knowledge: "When the massive magnitude and impact of phage predation were established at the birth of phage ecology in the early 1990s, predation became fixed—probably erroneously—in our minds as the dominant role of [the] phage.

Phages have become the killers of the sea, akin to their pathogenic viral cousins in hospitals."[12] This parallel or amalgam between bacteriophages and viruses with human hosts is widespread.

It is therefore important to understand how a certain conception of viruses, and how the terms used to describe them, might have posed an obstacle to research. The emphasis on predation and the conceptualization of viruses as dangerous or deadly deserve to be questioned—a delicate undertaking, to say the least, after two years of a global pandemic, but one that could shed new light on the latter.

Feminist science and technology studies have demonstrated that the language and metaphors used in scientific practices influence both the conceptualization of the problems encountered and the orientation of research programs.[13] In the case of bacteriophages, the language and metaphors are generally the same as those used for viruses or at least reuse certain characterizations. In *Contagious: Cultures, Carriers, and the Outbreak Narrative*, the researcher Priscilla Wald describes what she calls "outbreak narratives," techniques that epidemiologists can use to make transmission routes visible and help them anticipate and manage the course of a given epidemic. Using numerous examples, she shows that these narratives revolve around three moments: the emergence of an infection, its spread through a global network of contacts, and finally its containment or eradication. Far from being trivial, these narratives affect responses to the challenges posed by epidemics in a globalized world insofar as they stigmatize individuals and populations, as well as behaviors and lifestyles.[14] These performative narratives, which unfortunately have all-too-real consequences, share a certain number of constants, such as the regularly restated needs for barriers and for national and physical borders, the conception of "the other" as a permanent source of danger, the individualization of responsibility (as illustrated in the tracking down of a "patient zero" to whom all subsequent contagions are attributed), and the use of warlike metaphors (such as those used by the French President Emmanuel Macron when announcing the first COVID-19 lockdown on March 17, 2020).[15] They also perpetuate the idea that science is the fruit of the genius of a few "Great Men" (whereas all ethnographies of scientific practices underline the profound and necessarily collective nature of the production of knowledge and technologies), Great Men who, in the case of bacteriophages, are most often portrayed in the gender-stereotypical figure of the "virus hunter." This expression, coined by Paul de Kruif in 1922 primarily in the context of laboratory work, has subsequently been used with novelistic fervor in many researchers' narratives, setting the scene for a virile confrontation between "Man" and a radical otherness, most often conceived of as lurking in dark and oppressive caves, deep in the belly of an Earth-Nature perceived as threatening and hostile, for which reason it must be controlled and dominated. Scholars of feminist science and technology studies and ecofeminists have explored analogies between "Nature" and "Women," including Nature's (forced) unveiling, its treacherousness, and the need to control and exploit it.[16] Unfortunately, however, the narrative of a science born of the genius of a few Great Men still seems promised a bright future, as does the invisibilization of people and knowledge considered of lesser importance, not to say negligible or inferior.

However, the study of phages and bacteria calls into question the conception of a "predatory" virus by revealing the plasticity of these entities and their common existence within and via the relationships they establish. When a bacteriophage virus integrates its genetic material into that of a bacterium during a lysogenic cycle, the virus's mode of existence changes radically.

It becomes an integral part of the host bacterium. Some articles refer to such viruses as "dormant," which could give the impression that, in this specific configuration, the virus does nothing. Meanwhile, the bacterium in which the virus is "dormant" develops new abilities. To which entity do this abilities belong? To the bacteria or the virus? Do the two entities still exist independently of each other, or have they become an entirely new entity—a chimera? These questions are all the more complex because the new entity formed in this manner may have only a fleeting existence. Under certain conditions, the viral genetic material may separate—sometimes imperfectly—from the bacterial genetic material and be replicated hundreds of times to form new viral particles and then be released. Since lysogeny most often results in exchanges of genetic material, can we still consider the resulting entities to be the same virus, or the same bacterium?

Virologists who have developed in-depth knowledge of the world of phages and bacteria are reluctant to say that a phage "kills" its host. For example, Rémy Froissart, an evolutionary biologist with whom I have worked since beginning my research, told me that bacteria die when viruses have used up all of their resources. This reformulation, which may seem trivial, clears the way for an entirely different narrative, one in which viruses do not kill bacteria intentionally but rather use them as a matrix for a process of replication and creation that may sometimes include fragments of the bacteria themselves. This narrative puts the spotlight on creative powers freed from the dogmas of sexual reproduction and immunology and emancipated from the notions of self and nonself (or constantly creating and dismantling them)—just think of the transition from lysogeny to lysis. By exposing the porosity of the categories with which Western modernity apprehends and acts on the world, these organic intimacies reveal a more subtle spectrum of relationships between entities that are increasingly difficult to essentialize. To add to this complexity, these relationships also depend on the broader environments in which they exist and which they help shape and compose, sometimes with spectacular results, as the example of the lesser flamingos and the portrayals of the role of phages in biogeochemical cycles have shown us.

What is more, these interaction patterns are not unique to viruses and bacteria. In human DNA, 5 to 8 percent is of viral origin. In other words, it is the result of the integration—at a given moment in the evolutionary history of the species—of the genetic material of a virus with a host cell. The best-known example of this in the animal kingdom are syncytins, proteins required for placentation (the formation of the placenta, a capacity specific to mammals) that are derived from interactions with viruses.[17] This DNA of viral origin, a large proportion of which is considered "junk DNA" because of its supposed uselessness, constitutes the slow sedimentation of relational experiences, the precious archive of events that have brought mammals to where they are today.

The study of bacteriophage viruses and the analysis of some of the existing knowledge of retroviruses lead to a conceptualization of these entities that differs greatly from the conception lurking, sometimes benignly, behind the image of the "virus hunter." Viruses—relational entities with now seemingly boundless potentialities—are not "external" to other biological entities: They permeate them, shape them, and transform them in many ways; they work with them to create microgeohistories that blur or at least redefine the boundaries between certain categories. If an analogy must be found, it should be one of gathering— to reflect the differentiated activation of viruses' potentialities according to the type of container in which the activation takes place—rather than one of hunting, which permits a vision of viral relationships in just a single mode.[18]

FROM AMPHIBIOSIS TO PLURIBIOSIS

In 1962, the microbiologist Theodor Rosebury created the term "amphibiosis" to account for the dynamism of the relationships that can be established between two entities, depending on the context.[19] Martin Blaser, a researcher known for his work on human microbiota, then took up this idea in a book published in 2014. Learning about amphibiosis marked a turning point in my thinking about how to encapsulate, in a single term, the entire spectrum of relationships mapped out by phages and bacteria and, more broadly, by two living entities (e.g., humans and viruses). The term enables us to shift from describing the intrinsic properties of entities to describing the relationships established *between* these entities, relationships that are context dependent.[20] Essentializing these entities—freezing them in time or space to talk about "war" and "peace," "friends" and "enemies"—therefore constitutes an initial error and a first act of violence committed against relationships between living entities. This leaves only two alternatives: simple eradication or blind irenicism.

However, the notion of amphibiosis is reductive in two ways. First, as suggested by the very structure of the word (*amphi* + *bios*), it unintentionally perpetuates the idea of binary relationships.[21] Second, even if it enables us to embrace the dynamism of relationships, it is based on a relatively fixed conception of the entities involved and thus prevents us from considering how their relationships transform them. However, phages and bacteria can help us develop a decentered perspective based on the comparison of different temporalities: During one human life-time, thousands of generations of bacteria and billions of cells live and interact with bacteriophages, transforming one another and influencing the environments they inhabit. Since microbial entities evolve at a much faster rate than humans, they facilitate the observation of these evolutionary dynamics, the comprehension of which, in the animal and plant kingdoms, requires meticulous reconstructions linked to numerous assumptions.

These differences in scale reveal that the essentialization of entities and relations is a matter of time: It can be done only by stopping their movement (by placing the strains in the freezer at -80°C, for example) or if they are moving so slowly as to give an impression of relative immobility. Therefore, the idea of entities with a fixed essence maintaining univocal relationships in environments considered immutable can be explained by a significant difference in generation time: for humans, a few decades; for bacteria, sometimes only a few tens of minutes.[22]

I have chosen the term *pluribiosis* to describe the entanglement of multiple relational spectra between constantly developing entities, which are formed or transformed by their encounters with other living beings. Entities, relationships, and environments are inherently fluid and intensely relational. While essentializations and categories are required for scientific activity (and, more generally, for action), the notion of pluribiosis reminds us that they are situated instantiations or snapshots of the fundamentally relational nature of the living world. Reporting on the situations and relationships described and experienced by the many people—biologists, ecologists, physicians, pharmacists, regulators, and patients—whom I have met or read about has enabled me to apprehend the diversity and evolutions of these entities. The concept of pluribiosis is therefore intended not to replace concepts, metaphors, or terms used in various practices but rather to recognize this multiplicity, to embrace it and study its consequences in the various situations of knowledge production.

Pluribiosis is a heuristic concept because it recognizes the importance of relationships and environments, thus preventing us from making assumptions about what is of a biological or social, natural or cultural nature.[23] It pays particular attention to what becomes of the entities and to the transformations that shape the situations described and what they contain. It is also a prescriptive concept insofar as it acknowledges the fundamentally multiple and situated nature of knowledge concerning the living world (*pluri-* referring this time to the multiple ways in which scientific practices can approach a single subject), as well as the *transformative potential of this knowledge.*[24] If evolution and relationships are constitutive of the living world, then the knowledge that ostensibly accounts for them, and their uses, must be included in the configurations in which they are produced. In the case of phages, this means that we cannot "forget" how they were isolated and what their inclusion in a collection endorses: a snapshot of microgeohistories, a specific configuration, which could have been completely different, and which therefore cannot be easily removed from its context of production.[25]

This detour to examine the potentialities of phages, through what we currently know about their roles as gene shifters or composters of worlds, has not only helped us apprehend the concept of pluribiosis but also enabled the reintegration of a therapeutic practice into more global dynamics. Using phages as therapeutic entities means mobilizing only some of their potentialities, chiefly the capacity to lyse bacterial cells. This does not mean that these phages are devoid of any other capacity. On the contrary, although the phages in collections are strictly lytic, (i.e., supposedly incapable of performing lysogenic cycles), they remain relational, pluribiotic entities. What does this imply for the practice of phage therapy?

Phage collections are extremely valuable because they can be used to treat people and even save their lives. For this reason, they are coveted; like any health "good," they can be commodified and exploited. Pluribiosis draws attention to the creative powers of living beings while refusing to consider scientific knowledge as innocent. Reducing phages to the status of "snipers" or "professional killers" poses two risks: first, of depriving ourselves of many of their potential benefits and ignoring the fact that they may become precious allies; second, and concomitantly, of forgetting that these entities are recalcitrant and capable of doing other things—indeed, much more than what is expected of them.

Despite decades of practice, phage therapy is still in its infancy. This presents an opportunity: There is still time to consider the full potential of phages and bacteria and therefore to make choices. Because while everything is related, not all relationships are good. And if everything is related, then we must be aware that the various ways that phage collections are used will have various consequences. The following chapters will describe the choices ahead of us and their implications.

5

PLURIBIOTIC MEDICINE

What I see in the ICU is this: Enterobacter cloacae, Klebsiella, *coli* [Escherichia coli], *pyo* [Pseudomonas aeruginosa], Acinetobacter—*all bacteria. . . . Well, each one has its own specificities, you see, but* Pseudomonas *and* Acinetobacter *are bacteria that you don't see in the general population; they're opportunistic bacteria, which are not very virulent to begin with but which have a phenomenal capacity to acquire antibiotic-resistance genes and become competitive monsters, which, even if they're never very virulent, are extremely antibiotic resistant. Bacteria form biofilms, cling to everything, and then kill people slowly.* Klebsiella *are more or less the same: These are digestive germs that take advantage of the general state of debilitation in which some patients find themselves after undergoing surgery and reoperation, digestive procedures, etc.* Enterobacter cloacae *is no different. And episodically, you see* Staphylococcus aureus *infections, but* Staph *infections are more frequent in intensive care, in people with endocarditis who started out with a finger wound that became infected and spread. It's not the same kind of patient. So, depending on the type of patient, you have different types of infection, and each time it's different: each type of bacteria, each type of patient, each type of infection.*

PHAGES AND ANTIBIOTICS

Enterobacter, Klebsiella pneumoniae, Escherichia coli, Acinetobacter baumannii, Staphylococcus aureus, and *Pseudomonas aeruginosa*. Raphaëlle Delattre, a critical care physician at a large public hospital in the Paris region when I spoke with her in 2018, lists some of the bacterial species now considered the most dangerous by the World Health Organization. Some of these bacteria are highly pathogenic; others are much less so but have acquired this characteristic *opportunistically, owing to the environments* in which they evolved. These are bacteria that have become resistant to many, if not all, available antibiotics and thus cause chronic infections.

If phage therapy is enjoying a new lease of life in Western European and North American hospitals and laboratories after years of relegation to medical school archives, it is because these bacteria are resisting the increasingly bold attempts of infectious disease specialists and other experts to eliminate them from the bodies of people who come to them for help, sometimes in desperation. The current development of phage therapy can therefore be understood only in the context of antibiotic therapy (the gold standard for anti-infective agents) *and* antibiotic resistance (which it aims to remedy)—a fundamentally different context from that in which it emerged in early twentieth century, one in which it is now impossible to ignore the evolutionary capacities of microorganisms. In the opinion of the infectious disease specialists, internists, surgeons, and critical care physicians I encountered, phages are one of several alternative solutions to the problems posed by antibiotic-resistant bacteria.

However, although phages (evolutionary entities) and antibiotics (chemical molecules) seem to do the same thing, or are at least used by health care professionals in pursuit of the same objective, their modes of action and thus, presumably, their potentialities are extremely different. Drawing on the experience of hospital clinicians, this chapter examines how the modes of existence of phages are embodied in practices that convey specific conceptions of care, infection, and healing.

The principle of phage therapy is relatively straightforward, and its various stages are similar to those of antibiotic therapy: After isolating the bacterial strain(s) responsible for a patient's infection, one or more virulent phages active on the bacterial isolate(s) are identified via a phagogram (an in vitro test to study bacterial susceptibility to bacteriophages) and then administered to the patient in direct contact with the pathogenic bacterium or bacteria with the aim of triggering the lytic cycle. The phages infect bacterial cells to reproduce within them, and lysis causes bacterial populations to die out.

As discussed, the relationship between phages and bacteria is highly specific owing to the various stages of the lytic cycle, each being composed of mechanisms of recognition and exclusion and each providing opportunities to test the possibility of a relationship between phages and bacteria (see figure 0.2). Adsorption, the first stage of the encounter, depends on not only the ability of phages to bind to receptors on the external surface of bacteria but also the ability of viral nucleic acids to enter the bacterial cytoplasm. Replication depends on both the persistence of the same genetic matrix in the cytoplasm and its ability to replicate, to name but two factors. The capacities of phages at each stage then determine what is known as the "host spectrum," that is, the number and type of bacteria with which a given phage can enter into contact, as well as the phage's reproductive efficiency: the number of phages produced within a single bacterium.[1]

Antibiotics have two main types of effect: *bacteriostatic*, when their presence inhibits bacterial growth, and *bactericidal*, when their presence reduces the bacterial population. They target the fundamental molecular mechanisms present in major classes of bacteria. For example, β-lactam antibiotics (such as penicillin) inhibit the synthesis of bacterial cell walls, whereas quinolones inhibit DNA synthesis. In all cases, antibiotics never entirely eliminate a target bacterium; rather, they reduce the population sufficiently for it to be controlled by the immune system (if it is strong enough to do so at the time of infection). Although some antibiotics are referred to as "narrow spectrum," and new antibiotics are being developed to target more specific characteristics of a given bacterial species, these chemical molecules still have a relatively broad host spectrum (always broader than that of a bacteriophage) and therefore frequently cause the death of nontargeted bacterial species.[2]

As antibiotics are chemical molecules catabolized by the body (by the kidneys, liver, and immune system), they usually require repeated administration, sometimes on the same day. The degradation processes of these chemical molecules are generally toxic to the organs concerned (the liver and kidneys). The concentration at which and the length of time for which they can be applied without causing too much damage to the body are therefore limited. Although partially degraded by the immune system, bacteriophages benefit from in situ amplification by the very bacteria they target. The available data also indicates a lack of toxicity after bacteriophage administration.[3]

Antibiotics	Bacteriophages
Chemical molecules	Dynamic biological entities
Variable specificity (very low to medium)	High specificity
Degradation/catabolization by the body	Multiplication, exponential concentration within the infected site
Significant side effects	No side effects reported to date

These fundamental differences in the modes of existence and action of antibiotics (chemical molecules) and phages (biological entities) open up new therapeutic opportunities, subject to the availability of phages for treating sick people. Specifically, they could be used in situations in which antibiotics have proven ineffective, leading to infections that are no longer acute but chronic, that is, established. Such situations are becoming increasingly frequent, not least because of the increase in antibiotic resistance. In some cases, antibiotics are only partially able to reach the site of infection; sometimes, they are entirely unable to do so.

This particularly applies to complex osteoarticular infections. As Tristan Ferry, the deputy head of the infectious diseases department at Croix-Rousse Hospital in Lyon, points out:

Chronic osteoarticular infections are one of the most difficult infectious diseases to treat because there are many bacterial persistence mechanisms, particularly in the case of osteomyelitis, where dead bone and bacteria are present in the form of biofilm. And biofilm can also be found on the surface of implants, including osteosynthesis devices and prostheses. In fact, the problem is that we face different but related challenges: challenges for infectious disease specialists who have to consider how to eradicate the infection, which for them means explantation [removal of the prosthesis] in order to remove the biofilm because antibiotics are ineffective for this purpose; orthopedic surgeons, however, will consider function: If the patient can walk, then they will seek to avoid excessively altering this function. These two challenges are both related and, above all, opposed. To put it another way, if the aim is to cure the problem, for an infectious disease specialist, you either consider amputation, or you consider removing the prosthesis. But the problem is that you'll be altering the function. The advantage is that you will cure the problem because you are mechanically removing the biofilm—surgically—but this will alter the patient's function. If you leave the implant in place, you will perpetuate the infection. So, these two entities are linked, but they are sometimes incompatible. In other words, it is sometimes impossible to cure the patient without removing the prosthesis, unless we can find innovative treatments that can destroy the biofilm or allow an antibiotic to do its job.[4]

Bacteria have a complex social life. Each has a life of its own, and they can give birth to populations of millions of individuals in a matter of hours or days. When this reproduction occurs on a solid medium, bacteria have the capacity to form what are known as biofilms—complex multicellular communities in which the organisms differentiate to enable a genuine division of labor—and become capable of secreting an adhesive polymeric matrix (on any type of surface, including "nonstick" surfaces): a film-like coating made of various components that binds and protects them. Whereas bacteria were for a long time analyzed primarily as discrete entities—cells independent of one another and their environment—the cooperative behaviors of biofilms are now being increasingly studied, and the numbers of roles and functions that we are aware of are constantly growing as scientists observe and study them. Biofilms are fascinating communities; they can grow to extremely large sizes and are found in a prodigious variety of ecosystems: on rocks, tree leaves and trunks, the fur of certain animals, and in stagnant pools of water, as well as in pipes, on the surfaces of medical equipment, on mucous membranes, and on human bones. They form a protective shield against external aggressors such as chlorine and antibiotics, enabling bacteria to survive in hostile environments. This is another example of pluribiosis: Biofilms are constantly evolving, forming and transforming themselves with every new interaction and encounter, maintaining fluid relationships with their environment and with other species. The same applies to biofilms that develop through interactions with humans. Some are mutualistic or even symbiotic, whereas others may be pathogenic, such as the dental plaque responsible for cavities and some biofilms that develop in the bladder and cause recurrent urinary tract infections.

This is also the case for biofilms that develop on bones or prostheses, which are often unaffected by antibiotic treatment; thus, treatment with antibiotics would not only be ineffective but also potentially dangerous since repeated use may lead to the selection of antibiotic-resistant bacteria. The presence of biofilm causes infectious disease specialists to consider amputation or the removal of orthopedic appliances because they see no ambiguity: Healing is achieved only by completely eradicating the pathogenic bacteria and therefore by removing the biofilm. But Tristan Ferry's approach is more subtle in that it expresses a variety of perspectives, each conveying definitions of what *curing* and *providing care* really mean. Although eradication is the standard definition of *cure* from a microbiological standpoint, health care professionals sometimes consider the means used to achieve it outrageous since prosthesis removal and amputation are both major surgical procedures that cause pain and significantly affect patients' lives. For these reasons, such procedures are generally carried out as a last resort, sometimes after many years and multiple operations.

In response to the complexity of osteoarticular infections and the public health problem they represent, interregional reference centers for complex osteoarticular infections (centres de référence des infections ostéo-articulaires complexes [CRIO-ACs]) were established in France in 2008. These centers central-ize patient care and offer patients the benefits of pooled expertise and treatment covering the various dimensions of their condi-tions. Ferry, who heads the Lyon CRIOAC, conducts research on and promotes innovative treatments, including lysines, anti-biotic cement, and phages. Over the last five years, some thirty people have been treated with bacteriophage viruses at Croix-Rousse Hospital, mostly, though not exclusively, for osteoarticu-lar infections. Phages are thought to be capable of destroying biofilms in the human body, a capacity already proven in vitro (in test tubes) and ex vivo (on tissues placed in an artificial environ-ment outside the body) and which in certain cases could avoid the need to perform major, incapacitating surgery.

CHALLENGING THE DOGMA OF ERADICATION

The case of complex osteoarticular infections also sheds light on an assertion central in microbiology that *curing* means *erad-icating*, an argument that is becoming increasingly difficult to defend, especially when such a high price must be paid (add-ing another dimension to the expression "costing an arm and a leg"). Raphaëlle Delattre shared the following observation about chronic infections with me:

> I'm not the only one who thinks it's better to maintain a low-key colonization rather than trying to eradicate it, except that colo-nization is usually kept low-key only when it can't be eradicated. Instead, it could be used as a strategy from the outset. For peo-ple suffering from cystic fibrosis, associated with environments

that are conducive to bacterial proliferation, it's best to try control things rather than eradicate them—and to stop this current arms race in which we've created bacterial monsters. Because it really is an arms race, except that the bacteria are sure to win it. . . . I think we'll continue to use antibiotics in probabilistic, empirical circumstances. For example, when you arrive at hospital with pyelonephritis [a kidney infection], we're pressed for time, we don't know who you are, and we treat you with drugs because it's an emergency. But I believe that phages have their place in the treatment of more chronic infections. The problem as I see it is that antibiotics are constantly trying to eradicate things: niches that are actually favorable to strains. And the problem is, if you eradicate something that wasn't actually that bad, you don't know what might replace it.[5]

Pulmonary infections account for a substantial proportion of all bacterial infections worldwide, and are sometimes very difficult to treat because the composition of the lungs, consisting as they do of bronchi, bronchioles, and around three hundred million tiny pulmonary alveoli and having a total surface area of around one hundred square meters, that is, about fifty times the total surface area of the skin.[6] The lungs provide a humid environment particularly conducive to the proliferation of many kinds of bacteria and full of nooks and crannies that are sometimes difficult to access. As Raphaëlle Delattre points out, maintaining a low-key infection rather than striving for eradication could become a real strategy in the treatment of infectious diseases, one in which phages would have a role to play. This solution could avoid the need to administer multiple antibiotics whose effectiveness diminishes rapidly as bacteria adapt to them and which can cause a host of serious side effects such as liver and kidney damage, which is sometimes irreparable. This was the approach adopted several years ago by Marc, a young man with cystic fibrosis who, with the support of his brother, an internist, traveled to Georgia, where he could obtain phages that are inhaled in aerosol form, phages that are regularly adapted to account for bacterial evolution.

Eradication, central to the practice of infectiology, can become a dangerous game: In the end, "the bacteria are sure to win," thanks to their incredible adaptive capacities but also because an infection is never merely an unfortunate encounter between a bacterium and a fragment of mucous membrane in the lung of an unchanging body. The encounter takes place in a complex ecosystem in which the disappearance of one species may enable the development of another without any means of predicting how it will behave.

"If you eradicate something that wasn't actually that bad, you don't know what might replace it." These words sum up the reasoned and subtle position defended by many infectious disease specialists with whom I have come into contact over the last few years: Any entity can be truly apprehended and understood only through the relationships it maintains with its environment. Some bacteria defined as pathogenic may not be all that bad. Others, living in the same ecosystem and in the same human body, could prove far more damaging to their human hosts if the conditions allowed them to express their potentialities. In other words, the pathogenicity of a microorganism is contextual. However, there is nothing new about this ecological approach to infectious diseases. It was even explicitly supported by some of the most eminent microbiologists of the twentieth century (such as Theobald Smith, Frank Macfarlane Burnet, René Dubos, and Frank Fenner), but it remained relatively obscure until the 1980s, when it began to be used to explain the emergence of diseases and antibiotic resistance (a topic I will return to in chapter 8).[7]

Since the beginning of the twenty-first century, with the rise of metagenomics and the wealth of data now available on the human microbiota, this concept has become increasingly widespread. This is why a researcher working in the microbiology department at Croix-Rousse Hospital did not need to provide any further clarification during our interview when explaining that "the fact of eliminating a pathogen [with phages or antibiotics] modifies an ecological niche and, as a result, also has consequences for the microbiota." An ecosystem entity disappears, and the web of relationships is altered. Given the current state of knowledge, it is difficult to determine the consequences, so we need to be cautious.

CONSTITUENT RELATIONSHIPS

Antibiotic resistance and the creation of niches for increasingly pathogenic bacterial species are the most visible and worrying consequences of the massive use of antibiotics for the purpose of eradicating bacterial infections. However, most antibiotics are broad-spectrum molecules and therefore affect a wide range of bacterial species, including those that can now be described as commensal or mutualistic. While humans have always maintained constitutive relationships with the bacteria, viruses, protists, fungi, microscopic worms, and other forms of life that populate them, in the life sciences and human sciences, we are only just beginning to consider the consequences of these complex interactions.[8] In the medical field, however, this redefinition of organisms and their components has been accompanied by a reconsideration of the health conditions and etiologies of numerous pathologies in the light of the role that microorganisms might play in them.

So, what happens when a dose of antibiotics enters the body? Numerous microbe populations disappear. These disappearances, which are never total, are generally localized and most often cause inconveniences that almost everyone who has taken antibiotics at least once in their life has experienced—gastric disorders, mycosis, etc.—but which are transient and cease once the ecosystem (i.e., the intestinal flora) has more or less recovered. But what happens when the body is flooded with antibiotics, when our skin is constantly being cleansed with antimicrobials that we are not even aware of using? According to the microbiologist Martin Blaser, this leads to the definitive eradication of microbes, some of which, having coevolved with the human species—some for millennia—can justifiably be defined as companion species and some of which we need without yet knowing exactly why.[9] However, the historian Alexis Zimmer urges caution regarding accounts of disappearing microbes, claiming they are no more innocuous than accounts describing human–microbe relations in terms of war and eradication.[10] Instead of saying that microbes are disappearing, it might be more accurate, from an evolutionary standpoint and in the pluribiotic perspective adopted here, to consider that the conditions imposed by the increasing use of antibiotics have caused certain bacteria to evolve in such a way that they can no longer maintain relationships with the bodies that host them. Once again, the difference may seem slight, but this way of telling the story provides a very different take.

These changes in relationships are the correlate of antibiotic resistance: Superbugs are not the only product of excessive antibiotic use. Antibiotic resistance disrupts ancient, intimate, and constitutive modes of relationships, at least for humans. Indeed, a growing number of studies have presented evidence of both correlations and strong causalities between the absence of certain microbes and the onset of disease.

Thousands of years of coevolution seem to have resulted in microbes participating in the education of human immune systems. This is one explanation for the "hygiene hypothesis," first formulated in 1989 by the epidemiologist David Strachan and currently gaining in plausibility on a daily basis.[11] In the second half of the twentieth century, we witnessed a massive decline in infectious diseases thanks to antimicrobials, vaccination, the effectiveness of health care systems, and changes in lifestyle and hygiene practices. However, this decline was accompanied by an exponential rise in immune disorders such as type 1 diabetes, Crohn's disease, and multiple sclerosis, and allergic diseases like asthma (figure 5.1).

The hygiene hypothesis postulates a causal relationship between these two series of events, with a reduction in the frequency of infections contributing directly to an increase in the frequency of immune disorders and allergic diseases. Once again, the underlying mechanisms of this hypothesis have yet to be elucidated owing to the complexity of ecosystems and the modeling required, but numerous studies support the hypothesis, presenting the immune system as a system in dynamic equilibrium, regulated by both internal factors and the environment, particularly the microbial environment. The profound alteration in some of these factors, such as a reduction in the bacterial populations with which we are likely to interact, particularly from an early age, could therefore lead to an immunological imbalance triggering allergic or autoimmune pathologies.[12] This hypothesis seems to be corroborated by numerous experimental models in which viral or bacterial infections play a protective role against these same immune disorders.

Whether we adopt the hygiene hypothesis or Martin Blaser's "missing microbe" hypothesis, the absence of microbes seems to

Incidence of infectious diseases (%)

——————— Rheumatic fever
——————— Measles
- - - - - Mumps
- - - - - Tuberculosis
.... Hepatitis A

100
50
0

1950 1960 1970 1980 1990 2000

Incidence of immune disorders (%)

——————— Type 1 diabetes
——————— Crohn's disease
- - - - - Multiple sclerosis
- - - - - Asthma

400
300
200
100

1950 1960 1970 1980 1990 2000

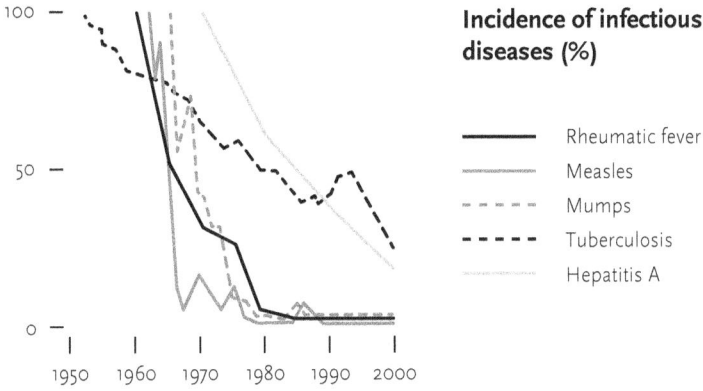

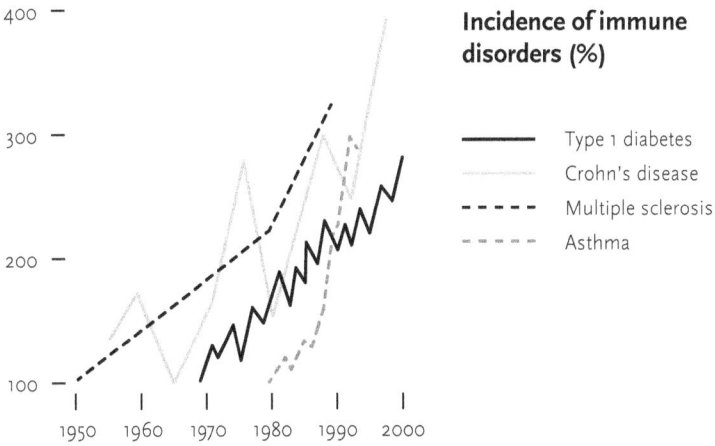

FIGURE 5.1 The inverse relationship between the incidence of prototypical infectious diseases and the incidence of immune disorders, 1950–2000.

Source: Adapted from Jean-François Bach, "The Effect of Infections on Susceptibility to Autoimmune and Allergic Diseases," *New England Journal of Medicine* 347, no. 12 (2002): 911–20.

be a cause of illness in humans. This reasoning may seem absurd, or counterintuitive at the very least. However, by moving away from a conception of disease as caused by discrete entities and instead focusing on the transformative relationships between humans and microbes (whose qualification as pathogenic or mutualistic is valid only in the context of transient relationships), it could be claimed that humans are getting sick because they are being cut off from their environment, enclosed in sanitized bubbles made possible by advances in chemistry and biomedicine.[13]

In an article titled "What Is a Pathogen?," Pierre-Olivier Méthot and Samuel Alizon show that the strict distinction between pathogenic and nonpathogenic agents is a legacy of the ontological model of disease. Moreover, this distinction ignores a number of details relevant to understanding pathogenesis processes, such as the various roles played by the host. Following in the footsteps of John Dupré, who developed a processual conception of life (as discussed in chapter 4), these researchers suggest that the biological phenomena at work in health and disease result from a "hierarchy of entangled processes, rather than a hierarchy of static things." We must therefore refrain from essentializing not only relationships but also entities of any kind, from microorganisms to humans. As we have seen with phages and bacteria, these entities are constantly transformed by encounters, occupying spectral positions in microgeohistories. They are pluribiotic embodiments: not friends or enemies, not parasites, commensals, or pure symbionts but spectra of intersecting and transforming relationships.

So, what does it mean to practice a form of medicine based on the eradication of microbes when our relationship with them appears to be far more complex and prolific than a one based purely on pathogenicity? This observation in no way means that we should not take action against microbial infections or that we should relinquish attempts at eradication altogether; that would be absurd. It simply means that a given phenomenon can and should be considered a snapshot of pluribiotic relationships and that, according to the many experts I encountered, we should pay attention to the microorganisms affecting human health and challenge the notion of eradication being the only appropriate response (or at least limit the eradication of bacterial populations to the strict minimum).

Phage therapy is a response to the exorbitant and varied costs of attempts at undifferentiated eradication, as well as to antibiotic resistance phenomena (which have also emerged partly, but only partly, as a consequence of the goal of eradication, as I will discuss in chapters 7 and 8). Unlike chemical antibiotic molecules, bacteriophages can target limited bacterial populations, thereby preserving the majority of the nonpathogenic bacteria present. Phage therapy could therefore offer the benefits of preserving microbial communities, whose role for humans and nonhuman living organisms alike remains unknown to the scientific community, and reducing the likelihood of resistance developing in untargeted bacteria.

Despite these undeniable advantages, phages are neither a miracle solution nor a replacement therapy. The many experts I spoke with kept repeating this in interviews, symposia, and workshops. These scientists and physicians work in a highly specific environment, tackling resistant bacteria and the lengthy, painful, and often incapacitating infections they cause, which are sometimes impossible to treat. It is precisely because they have been confronted with these recalcitrant bacteria and the long-suffering patients for whom amputation is often the only possible outcome that they are keen to preserve antibiotic treatment by using it more sparingly and more effectively. Confronted with the failures of chemotherapy, these experts have thus eschewed sweeping, one-size-fits-all solutions in favor of approaches that consider the specificity of each infectious situation. "Each time it's different," a critical care physician told me after describing the case of a patient who had contracted a lung infection immediately after receiving a lung transplant. "Each type of bacteria, each type of patient, each type of infection."

These assertions reflect an ecological and holistic vision of disease based on interspecific population dynamics but do not imply a rejection of eradicationism: Eradication is only one of several possible courses of action that can be taken to achieve cure, depending on the biographical context of the person to be treated. The distinction between acute and chronic infection is therefore fundamental, as is the distinction between phages and antibiotics. Two substances, two entities that, because of their differences, are not used to attain the same objectives. For the time being, antibiotics remain the best solution for acute infections, notably because of their broad spectrum, which allows them to be used without knowledge of the bacterium responsible for a given infection, thereby saving precious management time. Phages offer clear advantages for treating chronic infections, which are on the rise given the increase in antibiotic-resistant bacteria and the acquisition of antimicrobial resistance, which people with chronic infections often develop following repeated antibiotic treatments.

SUR-MESURE VERSUS PRÊT-À-PORTER

This is the highly specific context in which the 2011 paper "The Phage Therapy Paradigm: Prêt-à-Porter or Sur-Mesure," authored by nineteen internationally renowned researchers in the field of phage therapy, should be analyzed.[14] This is an opinion piece written after the first congress of the International Society for Viruses of Microorganisms was held at the Pasteur Institute in Paris in 2010, itself a sign of the growing community of experts dedicated to bacteriophage studies. In the article, the authors compare two conceptions of phage therapy: "sur-mesure" ("tailor-made") and "prêt-à-porter" ("off-the-rack"). After obtaining a phagogram, the sur-mesure strategy consists of administering only the phages active against the bacteria responsible for the patient's infection. The prêt-à-porter strategy, on the other hand, involves the development of "cocktails": assemblies of various phages active against a large number of isolates of a given bacterial species. These cocktails may contain up to several dozen phages, thereby providing significant "coverage" of the diversity specific to a given bacterial species, consequently mimicking the broad-spectrum effect of antibiotics.[15]

The authors are in favor of the sur-mesure approach, arguing that it would greatly limit the emergence of bacterial resistance. Because cocktails are "probabilistic" and do not specifically target pathogenic bacteria, they may be only marginally effective, and they may encourage the emergence of bacterial resistance, thus compromising the efficacy of future treatments. In line with the advances made in our understanding of the human microbiota in recent years, it can be added that by precisely targeting the bacteria responsible for a patient's infection, a less invasive sur-mesure approach could limit disruption to the microbial communities on which humans depend. Another argument in favor of a sur-mesure approach is also gaining popularity with the limited hindsight now available to infectious disease specialists: Although phages are not believed to trigger a strong immune response, they are recognized by the immune system. Thus, if a person who had undergone phage treatment once were subsequently treated with the same phages, their immune system might neutralize them before they come into contact with the intended bacteria, thereby reducing or eliminating their action. Another manifestation of pluribiosis, and of the plasticity of living beings, is the fact that we never witness the same encounter twice.

However, the need to write a paper like "The Phage Paradigm," which has been mentioned regularly at workshops and symposia since its publication, is a clear indication that the legitimacy of a sur-mesure approach to phage therapy is far from assured. The need for a treatment that exploits the pluribiotic capacities of living beings is regularly put forward to challenge the preeminence of the prêt-à-porter approach. While this approach relies on the selection of a handful of particularly effective phages from existing collections and their combination in the form of ready-to-use cocktails, the sur-mesure approach requires the creation of sufficiently large and varied phage collections to cover the plurality of pathogenic bacterial strains. Given the number and diversity of bacteriophages on our planet, particularly in the environments in which the targeted bacteria can be found, such as fecal matter and wastewater, such collections could be fairly easy to create (as we have seen, some already exist in research laboratories, which provide a valuable reservoir).

What is more, the sur-mesure approach already exists in the practices of hospital clinicians—and in the desirable futures of a significant proportion of the community.

It is the maintenance and even the existence of the prêt-à-porter approach that raises questions. This is an approach that turns phages into ersatz antibiotics, which could therefore be used in a similar manner according to existing practices and protocols. It is an approach that makes something old out of something new, exposing various communities to already well-documented risks. In fact, this approach denies what makes bacteriophage viruses special: the pluribiotic capacities they share with bacteria.

To understand the importance of increasing awareness of an approach the scientific and clinical foundations of which are based on common sense—and are defended by the vast majority of clinicians—we must link the development of phage therapy more closely to existing practices, standards, rules, and infrastructures.

6

EVIDENCE-BASED MEDICINE

The venerable age of phage therapy is regularly mentioned when discussing the thorny issue of demonstrating the efficacy of bacteriophage viruses for the treatment of infections—but in two contrasting ways. On one hand comes the fairly traditional mantra of the detractors (who are constantly declining in number, I should emphasize, having witnessed an exponential increase in support for phages since the start of this research): the pithy, "If it worked, we'd have known by now." On the other hand, and symmetrically, comes the reaction of those who have ardently campaigned for the recognition of phages as a therapy, some for many years: "Evidence? Phages have been saving lives for a hundred years! We've got plenty of evidence!" The question of evidence is also regularly raised in papers written by promoters of the sur-mesure ("tailor-made") use of phages, most often to emphasize that, owing to the specificity of these entities, their efficacy cannot be demonstrated by clinical trials.

These positions bear witness to the hegemony of *evidence-based medicine*, which some of its proponents define as "the conscientious, explicit and judicious use of current best evidence in making decisions about the care of individual patients, . . .

integrating individual clinical expertise with the best available external clinical evidence from systematic research."[1] The evidence-based medicine movement, which is highly codified after decades of development, is based on two complementary postulates: universal biology and universal rationality.[2] These conditions are sine qua non for the aggregation and generalization of results produced in temporally and geographically disparate conditions, which often concern extremely heterogeneous populations whose diversity is expressed in the form of standard deviations. The cornerstone of this system is the *randomized controlled trial*: the mandatory step for any biomedical therapy or intervention for which marketing authorization is sought and which stabilized in form toward the end of the 1950s.[3]

This is the backdrop against which we must consider the positions adopted by scientists and health care professionals who see clinical trials as an unsuitable means of evaluating a therapy intended to be sur-mesure and precisely situated, a therapy guided by the notion that each infection—and each potential cure—is a unique event. This chapter will focus on the tensions between, on one hand, the normalization and standardization required for the evaluation and existence of a therapy and, on the other hand, the preservation of the originality of a sur-mesure approach, that is, one that considers microgeohistories. How do clinical trials make (or break) phage therapy?

The question of how to assess the efficacy of phages has been posed almost since we started using them. The various literature reviews published by the prestigious *Journal of the American Medical Association* in 1934, 1941, and 1945, for example, highlight the controversies and lack of knowledge of the nature of phages and their mode of action, as well as the inadequate proof of their efficacy.[4] These reviews contributed to the marginalization of phages at a time when purportedly easier-to-use antibacterials (sulfonamides, followed by antibiotics, as I will discuss in chapter 7) were entering the market.

Few traces remain of phage therapy practices in the second half of the twentieth century (i.e., after the introduction of antibiotics to treat infectious diseases); of those that do, most are case studies and series reconstructions, and most of the data lies dormant in the archives of centers that have been closed or even destroyed.[5] Once phages had been removed from pharmacopoeias, the use of preparations from Russia and Georgia to treat people who had failed to respond to treatment with antibiotics was deliberately not reported.[6]

Although the question of phage efficacy remains a sensitive issue today, the scientific and clinical context in which phage therapy has been developing over the last fifteen years has little (if anything) to do with the environment in which it emerged over a hundred years ago or with the circumstances in which it virtually disappeared from Western Europe and North America in the 1940s. Over the past century, knowledge has developed and societies have changed: Phages have been recognized as viral entities; antibiotics have become mass-produced and mass-consumed; antimicrobial resistance phenomena have multiplied; the cost of the pursuit of eradication has become exorbitant and therefore more visible; microbiota have become a new object of knowledge; ecological awareness has developed; and the forms of evidence production and administration have changed. In other words, the long-term data available on the medical use of phages is far from proportional to the time that phage therapy has existed. This brings us back to chapter 2, where I mentioned that phages have been removed from the *Vidal* pharmaceutical dictionary. To reinstate them, it needs to be proven that they belong there.

CREATING AND EVALUATING A
THERAPEUTIC SUBSTANCE

A new therapy is generally developed in four phases, preceded by a "pre-clinical" phase consisting of a series of in vitro and then in vivo studies on animal models.[7] The phase I trial assesses the potential toxicity and the safety of the therapy in a very small number of people and study the kinetics and metabolism of the substance in the human body. The phase II trial (the "pilot" phase) provides an initial measurement of efficacy while determining the most appropriate dosage for a relatively small number of people. The phase III trial (known as the "pivotal study") is a comparative study of the efficacy of the substance to be evaluated versus that of either the reference treatment (i.e., the current standard of care for the pathology in question) or a placebo (if there is no existing treatment for the pathology in question). If the results are conclusive, this study can be used to support a marketing authorization application. Last, a phase IV trial studies the long-term occurrence of complications and delayed adverse effects.

These various phases, phase III in particular, have been studied extensively.[8] What is interesting about phase III is that it does more than simply validate a supposed causal relationship (e.g., between a substance and cure); it also establishes a framework within which that causal relationship is posited and constructs the entire proposition. To this end, the phase III trial, like any experimental method, proceeds by reducing the environment: the inclusion criteria chosen, the indicators selected to monitor the participants, and the types of questions asked are all designed to neutralize the majority of idiosyncrasies, which are relegated to the level of background noise. This renders the various entities of the trial quantifiable and, above all, mutually commensurable. The participants—reduced to their body or a part of their body—can be compared beyond their personal trajectories, and the effects of the treatment being evaluated—objectified via a series of measurements that depend on the proper standardization of instruments, reagents, sampling kits, professional routines, and specific forms of division of labor—can be aggregated. The trial therefore defines the substance, the disease, the criteria for cure, and the conditions of use, but also behaviors and lifestyle habits. Phase III enables a major ontological change in which a candidate substance becomes a drug.[9] In this way, randomized controlled trials can be said to *produce* therapies as much as they *evaluate* them.[10]

However, the randomized and controlled aspects of these types of studies have another consequence, as highlighted by Joe Dumit and Emilia Sanabria: They enable the healing power to be attributed to the molecule alone, making it a "magic bullet," as the popular expression goes. Dumit and Sanabria also point to the patriarchal structure of a narrative that elevates the drug to heroic status and renders the work that goes into medical, psychological, emotional, and collective care invisible.[11] Using the example of the evaluation of ayahuasca as a psychedelic substance, they point out that randomized controlled trials take on a distinctly colonial dimension when the therapies to be evaluated are not based on the premises of Western biomedicine. In this way, they "contribute to the erasure of the ongoing destruction of indigenous communities and territories and extraction/perversion of their knowledge systems."[12]

Randomized controlled trials are powerful tools for those who know how to use them. For example, numerous historical and anthropological studies have analyzed how such trials have been used to serve an ever-increasing capitalization of people's health. Forging the notion of "surplus health," modeled on Karl

Marx's concept of "surplus value," Joe Dumit shows how efforts to produce evidence, combined with profound changes in the notions of care, illness, and risk, have created a conception of health that has been appropriated and is appropriable by capital.[13] The objectives of randomized controlled trials underwent profound changes in the second half of the twentieth century. The initial aim, in response to the exponential increase in treatments marketed by pharmaceutical manufacturers (notably antibiotics, as I will discuss in chapter 7), was to establish codified procedures for assessing the safety and efficacy of purported remedies to safeguard the health of populations, as well as to establish an ideal of social justice.[14] However, in a singular turn of events, randomized controlled trials have also become machines for churning out new treatments and prescriptions.[15] They are managed as investments with the aim of increasing yields and ensuring healthy returns on investment, including by relocating them to low- and middle-income countries.[16] Each validated trial may lead to the marketing of a new molecule, the extension of the use of an existing molecule to a new medical condition or disease, or to an increase in the number of people eligible to take an existing molecule. A molecule whose development cost will have been all the lower, in the most extreme (but far from the rarest) cases, for having been based on the invisibilization of the work required to change its ontological status: the work performed by the people recruited as participants for the trial and the work carried out by the health care and support staff involved in the trial, often under precarious contractual arrangements and under difficult conditions.

This is not an innocent detail; clinical trials are not innocent devices. How they are constructed; the questions they ask; the criteria, norms, and standards they produce—all convey propositions and ways of establishing the existence of certain relationships or ignoring the existence of others, of attaching importance to certain beings or rendering them invisible. While these studies may be central to the pharmaceuticalization of human societies and health-based capitalism, they can also provide patients recruited as study participants with access to high-quality care in which they can play an active role.[17] It all depends on the type of moral and political proposals that researchers want to put forward or, as Isabelle Stengers puts it, "to make matter." Bacteriophage viruses are no exception to this rule, which explains why the implementation of clinical trials is the key to the development of phage therapy.

THE EUROPEAN PHAGOBURN TRIAL

PhagoBurn, the first large-scale clinical trial on bacteriophage viruses, was the fruit of a partnership between public bodies and a start-up firm called Pherecydes Pharma.

In 2008, the World Health Organization organized a working group to study battlefield burn victims, of which Patrick Jault, then a military doctor at the Percy Military Teaching Hospital in Clamart, France, was a member. Serious burn victims face numerous health issues such as an increased risk of sepsis owing to the loss of the skin barrier and immune deficiency, which makes them more vulnerable to infections. At the same time, Pherecydes Pharma was isolating and characterizing phages from wastewater. For this reason, Jault invited Jérôme Gabard, then the managing director at Pherecydes, to a NATO meeting. In a context of growing antibiotic resistance, the meeting was held to convince stakeholders of the need to conduct a clinical trial to assess the effectiveness of bacteriophage viruses in treating bacterial infections in patients with serious burns. However, phages have been excluded from European pharmacopoeias for several decades and have no regulatory status as a therapeutic product. Many discussions were then held by various regulatory agencies, notably the European Medicines Agency and the National Agency for the Safety of Medicines and Health Products (ANSM), during which a consensus emerged on the status of phages as medicines to determine the type of trial and protocol to be put in place (e.g., inclusion and exclusion criteria, number of participants to be recruited, primary and secondary outcome measures, adverse effects, monitoring, administration).

The PhagoBurn trial finally began in July 2015.[18] Its aim was to compare the efficacy and tolerability of a cocktail of lytic bacteriophages used to treat *Pseudomonas aeruginosa* with the current standard of care: an antibiotic cream containing 1 percent silver sulfadiazine that acts by blocking the bacteria's production of folic acid. Between July 2015 and January 2017, twenty-seven people over the age of eighteen years with a confirmed *Pseudomonas aeruginosa* infection in burn wounds were recruited from nine centers in France and Belgium and then randomly assigned to one of two arms of the trial: phage-based or antibiotic-based treatment. Each treatment was administered locally daily for seven days, and there was a fourteen-day follow-up.[19] A reduction in bacterial burden following treatment was the trial's primary outcome measure, that is, the key element used to judge the efficacy of the treatments being compared. If the reduction of bacteria in the wounds of participants in the intervention arm (i.e., those treated with phages) was shown to be statistically significantly greater than for participants in the control arm (i.e., those treated with antibiotics), then the greater efficacy of phages would have been demonstrated. But that is not how the story unfolded.

The findings of the PhagoBurn trial—that phages "reduce the bacterial burden in wounds at a slower pace than standard of care"—were considered disappointing by many in the phage community, especially given the substantial amount of money (several million euros) invested and the hopes raised.[20] There are several explanations for the study's results. First, the relatively low number of participants resulted partly from the difficulty of finding people with monobacterial infections since burn wound infections are predominantly polymicrobial. Second, the concentrations of bacteriophages in the cocktails administered to some participants were later found to be *a million times lower* than at the time of production because of the inhibition of certain types of bacteriophages when stored together, an aspect of phage preparation that was previously unknown. This meant that some participants in the intervention arm received a much smaller quantity of phages than specified by the trial protocol— indeed, it was virtually zero. This point is important to acknowledge because it enables the lack of efficacy to be attributed not to the phages but merely to their *absence*. Further, a post-trial study of microbiological samples taken from the participants showed that some of the bacteria isolated from them were resistant to the bacteriophages in the cocktails administered to them. Thus, even if the phages had been present in sufficient quantities, they would still have been unsuitable given the participants' bacterial strains.

However disappointing its results, this trial in no way invalidates the efficacy of phages. And, although this is minor compensation in view of the sums invested and hopes raised, it has enabled certain parties—including those with and without stakes in the trial—to learn valuable lessons, especially with regard to the production and storage of viruses, which are crucial to the development of phage therapy.[21] It has also encouraged them to reflect on the indications for which phage treatment might be relevant, as well as on appropriate therapeutic regimens.

The PhagoBurn trial revealed the immensity of the task that lies ahead: Although phages seem to do the same job as antibiotics, their mode of action is completely different. As freely admitted by the experts I have met over the last few years, whether researchers, clinicians, or regulatory agency employees, everything must be reconsidered or redesigned from scratch, including the types of infections to be treated; the techniques used to define phage activity (phagograms); the number of phages to be used; the dosage (according to the pharmacodynamics and pharmacokinetics, which must also be analyzed); the mode of administration (e.g., cutaneous, intrathecal, intravenous); the equipment to be used and the materials used to manufacture it (because phages have the ability to adhere to nearly any type of plastic); the premises in which the preparations are produced; the production standards; the efficacy measurements; and the evaluation criteria.

However, choices must also be made regarding the phage use strategy: Should cocktails be tested on participants? Or should study participants be administered only those phages that have already been shown to be active in vitro (i.e., by a phagogram) on the strain or strains responsible for their infection?

FROM COMPASSIONATE USE CASES TO CLINICAL TRIALS

Although, from a regulatory standpoint, proof of phage efficacy is a prerequisite for obtaining marketing authorization, many of the experts I encountered use this evidentiary constraint as an opportunity to establish certain rules for the use of phages; as they see it, if no clear proof of phage efficacy has yet been obtained, it is above all because they are being misused. This brings us back to the normative dimension of clinical trials, which are a means of not only evaluating the efficacy of a treatment but also *creating* the treatment. These carefully designed tests are not limited to providing a binary "it works/it does not work" type of answer; they also provide opportunities to develop new, fundamental, practical, and technical knowledge. However, this is an extremely difficult exercise as the developments of recent years prove.

In this context, compassionate use cases offer not only an alternative to people who have failed to respond to treatment but also an opportunity to test the potential of phages, refine their modes of use, and promote their development. This exposure is vital: Start-ups need it to attract and retain investors, and public-sector laboratories need it to obtain funding for the numerous calls for projects on which their research depends. These cases are a way of making progress and showing that something important is happening, but they do not provide the "proof" required in evidence-based medicine. They can be seen as microgeohistories: encounters in time and space among sick people, the pathogenic bacteria they carry, and the phages administered to them. In each case, it is a question of determining the best possible treatment plan for a patient and their care-givers based on the specific history of their infection and their wishes. What bacterial species is or are making them sick? Since when? Where are they located? Are the bacterial species always the same? Have they changed over time? What type of resistance have they acquired through the various antibiotic treatments taken? What is the state of the person's liver and kidneys? If it is a bone infection, can the bone be saved? What procedures can be used? Questions like these are important when determining the feasibility of administering phage therapy and the appropriate treatment regimen, but they are also crucial when assessing the efficacy of a chosen treatment regimen. In this regard, compassionate use cases provide an opportunity to explore the best conditions for bringing phages and bacteria together, this time no longer in the relatively simple and stable environment of Julie's Petri dishes and Eppendorf tubes but within the bodies of infected individuals, in complex environments and ecosystems about which scientists know very little. Each new case is an opportunity to learn and improve our limited understanding of the spectra of relationships in which health care professionals must intervene. Over time, physicians will come to regard certain factors as important and others as insignificant. They will try to detect regularities and hypothesize about the most favorable conditions for phages to reach bacteria and trigger a succession of lytic cycles with them while limiting opportunities for that bacteria to bring their adaptive capacities into play. In short, doctors will manipulate microgeohistories.

Because of current regulations, however, all compassionate use cases treated are by definition complex cases for which phage therapy is seen as a last resort.[22] And because of the multiple treatments taken by patients (many of whom also suffer from other pathologies, such as cancer), it is extremely difficult, if not impossible, to attribute the cure of a patient's infection to the administration of bacteriophage viruses. This is why, in the opinion of many of the actors I encountered, clinical trials are still necessary. By going beyond the strict context of compassionate use, they could recruit participants with less severe, less complex conditions, among whom commensurability would be easier to establish.

As Tristan Ferry explains, "All the experience we've gained in treating these people enables us to formulate hypotheses and design the best possible future clinical trials." For this infectious disease specialist, designing the "best possible future clinical trials" means establishing a clinical trial protocol that considers the specificities of bacteriophage viruses and the idiosyncrasies specific to the microgeohistories of each participant recruited while ensuring that indispensable commensurability.

This is where the tension between normalization and standardization, on one hand, and the need to consider the situated nature of phage therapy, on the other, comes into play—a tension of which Ferry, like the other infectious diseases specialists I met, is perfectly aware. While their primacy can be partly explained by the immensely important role they have played in the development of Big Pharma in recent decades, clinical trials are nonetheless experimental methods. As such, they can be configured to correspond as closely as possible to the question that needs to be asked. In other words, while clinical trials have mostly become machinery for producing universal elements, there is nothing to prevent the design of trials that account for specificities, as in the case of "n-of-1" trials, which involve just a single participant and are used to evaluate what are referred to as "personalized therapies," that is, treatments tailored to each patient's unique characteristics (such as their genetic make-up); this type of trial is commonly conducted in oncology. The problem of clinical trials being ill suited to the specificity of phages is therefore not linked to the method itself but rather to how the method is constructed.

YET ANOTHER ANTIBIOTIC

The PhagoBurn trial can teach us another lesson. As we have seen, regulatory agencies consensually decided to consider phages medicinal products when the study was being developed, but it was these same agencies, primarily the ANSM, that

imposed the phage cocktail model and did so for one simple reason: to ensure the comparability of phages with antibiotics in
terms of host range and therefore, potentially, of efficacy.

Because they can be used as anti-infectives, phages are lik-
ened to antibiotics, even by the most active promoters of their
development. "Phages are one solution among many because
there are lots of interesting ideas emerging. There are new anti-
biofilm antibiotics, antibodies combined with antibiotics that
deliver antibiotics inside cells and specific anti-infective gels that
can transport antibiotics," explains Tristan Ferry. This assimila-
tion is described in even clearer terms by Bérénice, a physician
who was developing a trial protocol when we spoke: "I see it
more as a toolbox: What tools do I have as an infectious disease
specialist when I'm confronted with an infection, and what do
I use to treat that infection? Phage therapy is one of these tools,
but I see it as yet another antibiotic. . . . It is an additional tool
you can use to treat the patient."[23]

This interpretation has been reinforced by the latest European Medicines Agency guideline concerning the evaluation of
drugs indicated for the treatment of bacterial infections. The
guideline, published in 2019, states, "To facilitate clinical development programmes for new antibacterial agents . . . there is a
need to ensure that each clinical trial carried out can be designed
to meet the requirements of multiple regulatory agencies."[24]
While phages were not mentioned in previous versions, the 2019
guideline states that its principles also apply to the development of phages. As Bérénice points out, "This means that the
approach may differ, but, in every case, we need to adhere as
closely as possible to the existing guidelines for antibiotics."[25]

This new guideline implies that the efficacy of phages should
be proven in the same way as for any antibiotic. Because of ethical standards stipulating that participants in clinical trials should

always receive the best available treatment, phages are almost always administered alongside antibiotics to give participants the best chance of recovery. This point has a major impact on the design of clinical trials, as highlighted by Tristan Ferry: "To obtain indisputable scientific proof, you need to carry out a therapeutic trial, a therapeutic trial with hundreds of patients who are randomly selected for surgery with antibiotic treatment versus—and this is where it gets complicated—other patients with the same infection who receive surgery with antibiotics *and* with phages."

This means comparing participants in a control arm (i.e., those receiving the current reference treatment, antibiotics) with those in an intervention arm (i.e., those receiving both antibiotics and phages). What is being evaluated is not one type of anti-infective (antibiotics) versus another (phages) but only the potential increase in efficacy associated with the addition of phages to antibiotics. The consequences of this are far from negligible. First, because phages are specific to a given bacterium, all participants recruited must be infected with the same bacterium or even with just a particular strain of a bacterium. Second, while antibiotic resistance is an existing problem, it is also a problem in the making insofar as antibiotics still retain much of their efficacy. Consequently, a very large number of participants would need to be recruited to show a statistically significant increase in antibiotic efficacy with the addition of phages. This is what makes clinical trials so long and costly.

Another consequence of equating phages with antibiotics in controlled clinical trials concerns the chosen outcome measure. In light of the protocols and discussions held with investigators of current or future clinical trials, "a robust outcome measure for an anti-infective product generally shows that the infection no longer exists, that the bacteria are no longer present."[26] However, the current use of phage therapy makes this criterion difficult to document. For example, for two trials that were about to begin at the time of writing, one on *Staphylococcus aureus* infections of prostheses and the other on foot ulcer infections in people with diabetes, it will be difficult to document eradication of the bacteria once the wounds have healed. Eradication as an outcome measure also raises questions about the relevance of definitions of *cure* in the context of infectious diseases. Indeed, as we saw in chapter 5, although the goal of eradication is justified in certain cases—notably in the treatment of acute infections—my interlocutors are more circumspect with regard to the need for eradication in chronic infections, for which phages are considered most appropriate.

It is essential to prove the efficacy of phages. But it is now easier to understand why some of the people I met view clinical trials as a hindrance to this goal at best and a totally unsuitable experimental method at worst. The problem is not so much the nature of phage therapy itself—a medicine based on highly specific evolving entities—as the ontological and epistemological framework within which it is situated. Clinical trials create therapies. The conceptualization of antibiotics, antibiotic therapeutic regimens, and the underlying conception of disease contribute to a set of norms and standards embodied in the very notion of antibiotics. Likening phages—dynamic and evolving entities with a specific mode of action—to yet another antibiotic amounts to imposing a definition of not only what an anti-infective treatment should be and how it should act (ontology) but also the methodology designed to evaluate it (epistemology).

Consequently, the question is not about whether it will be possible to evaluate phage efficacy (clinical trials are sufficiently flexible and adaptable) but rather what will actually be evaluated: Pluribiotic entities? Ersatz chemical molecules? A sur-mesure approach? A prêt-à-porter ("off-the-rack") approach?

The influence of antibiotics is overpowering, as Bérénice points out: "I'm the first to say that the dogma of antibiotics has a hold over me. There are guidelines, and we have to follow them as closely as possible, or it won't be accepted. Proof of efficacy will always be required. If today's blueprint is no longer valid, which blueprint will we follow? What is acceptable? I don't have the answer, and I can't find it on my own. Perhaps we need to completely rewrite the story, the whole story, to provide proof of efficacy. And we won't be doing it alone." Today, the "dogma of antibiotics" is one of the major obstacles to the development of a therapy that could provide (admittedly partial) responses to the problems that antibiotics have helped create—as part of a sur-mesure approach rather than as an umpteenth antibiotic. This is all the more relevant since antibiotics are not the "magic bullets" of twentieth-century medicine—far from it. To understand the key issues surrounding the current development of phages, we must take a closer look at these chemical molecules. To "completely rewrite the story," we must first understand every detail of the story.

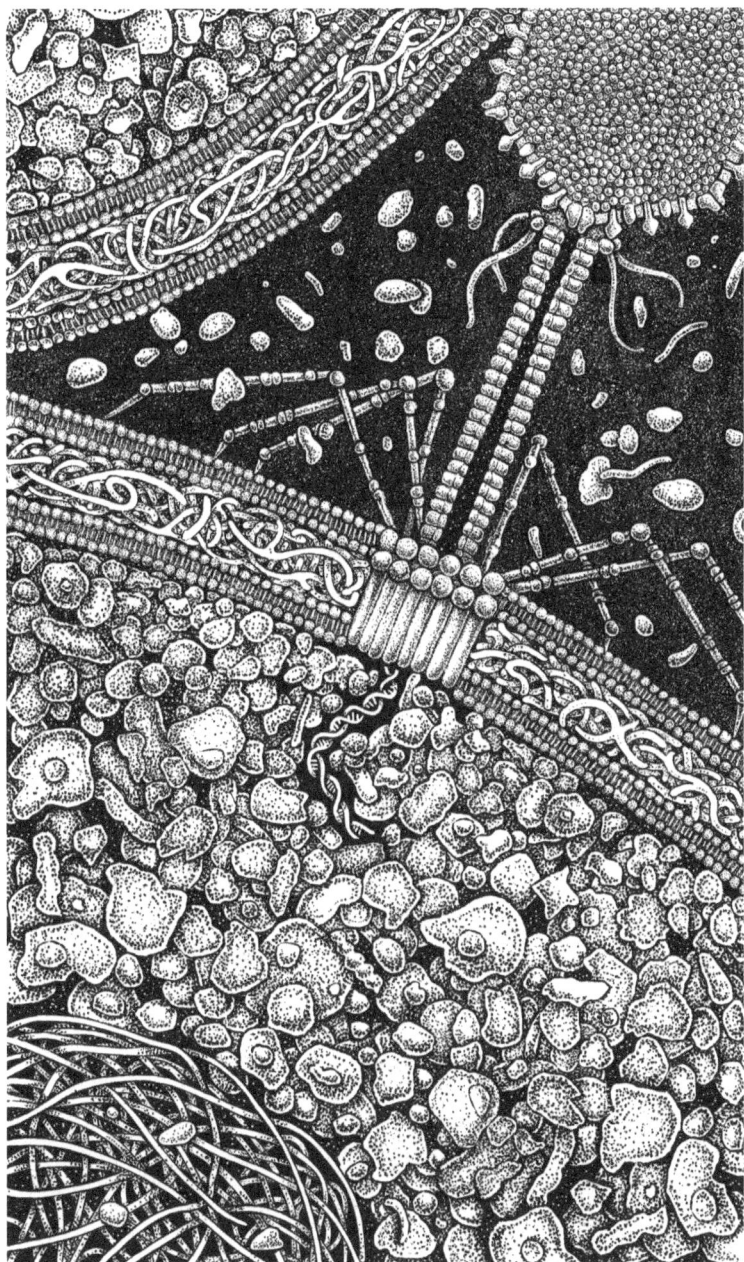

FIGURE 6.1 Pluribiosis

7

ANTIBIOTIC INFRASTRUCTURES

I n a 2019 paper, the anthropologist Clare Chandler suggested that we think of antibiotics as infrastructures according to Susan Leigh Star's definition of the word: not as simple physical, material systems but as complex, structured systems of relationships between living beings (human and nonhuman), things, and discourses.[1] While Chandler focused almost exclusively on the programmatic stage, I shall look back briefly in this chapter on the history of antibiotics and the formation of antibiotic infrastructures, as well as on the production and use of these molecules and the knowledge accompanying their use, in an attempt to explain their density, their pervasiveness, and the fact that they are impossible to avoid.

Whether phage therapy is presented as an alternative or complement to antibiotic therapy, whether the possible uses of phages are contemplated in human or animal health or in the environment, it is impossible to consider their development seriously without situating them in the context of the antibiotic infrastructures that we have inherited, for better or for worse.[2]

A BRIEF HISTORY OF ANTIBIOTICS

The established account of the discovery of penicillin is an excellent example of the history of "Great Men" (discussed in chapter 4) in which an often erratic assembly of humans, objects, devices, microbes, regulations, instruments, methods, and policies is overshadowed by the ideological power of a discourse that has established Alexander Fleming alone as the person who gave the world the first antibiotic. However, the history of antibiotics is a much more complex story that merits a closer look.

In September 1928, Alexander Fleming returned from vacation to his laboratory at St. Mary's Hospital in London to resume experiments on the action of an antibacterial enzyme on staphylococci. As frequently happens in laboratories, he soon discovered that his Petri dishes had been contaminated by another microorganism, a fungus called *Penicillium notatum*, on which his lab neighbor, a young Irish mycologist, was working. The story might have ended there if Fleming had simply disposed of the dishes and embarked on a major cleanup. However, the microbiologist's trained eye spotted the presence of a circular zone around the mold in which there were no staphylococci. He concluded that the fungus, which he named "penicillin," was capable of secreting a substance with antibacterial properties.

However, as Wai Chen points out in *The Laboratory as Business: Sir Almroth Wright's Vaccine Program and the Construction of Penicillin*, in February 1929, "the word 'penicillin' implied nothing more than the formula 'filtrate of mold-fluid culture' and the rest of its properties (including those we know today) were developed later."[3] In the following years, Fleming used penicillin mainly as a reagent to identify and discriminate between bacteria. It was not until the end of the 1930s that the biochemist

Ernst Chain, the pathologist Howard Walter Florey, and the biologist Norman Heatley, with the help of their teams, succeeded in purifying penicillin and using it on infected people. The large-scale production of this first antibiotic would subsequently be made possible by the departure of Chain and Florey for the United States. For several years, the team at the Northern Regional Research Laboratory in Illinois focused on improving production techniques, whether by developing suitable culture media or by searching for *Penicillium* strains with a better capacity to synthesize penicillin. One of the secretaries in the laboratory—who was also a trained microbiologist—was apparently responsible for the discovery of *Penicillium chrysogenum*, the strain that would eventually be used for production.[4] The team relied on the existence of specific techniques, first and foremost a stirred-tank fermentation technology developed in 1929 by Pfizer, a company then specializing in the production and refining of chemicals.[5] In 1944, production was sufficient to meet the needs of the US armed forces (figure 7.1).[6] In 1945, penicillin was made available to the American population.

Penicillin ushered in the era of "wonder drugs," drastically reducing mortality—deaths by septicemia and pneumonia, to name but two examples—and significantly improving conditions for reproduction, especially by enabling the treatment of both congenital syphilis (responsible for miscarriages and malformations) and gonorrhea. It was soon joined by streptomycin, with these two molecules alone accounting for 99.7 percent of antibiotic production in the United States in 1948. Very quickly, however, the various pharmaceutical companies began searching for new molecules for two key reasons. First, the application spectrum of these two antibiotics was relatively limited considering the promise of this new type of anti-infective agent. Second, because of their development model, these molecules were

FIGURE 7.1 A promotional poster created by the US government
between 1942 and 1945

not covered by an exclusive patent, thus offering little guarantee of profitability. This is what led the pharmaceutical company Lederle to isolate and then produce aureomycin, an antibiotic molecule with a broad spectrum of activity (i.e., activity against many bacterial species) approved by the Food and Drug Administration (FDA) in December 1948. Parke-Davis followed suit with chloromycetin in 1949.

For the historian Scott Podolsky, however, the real change came when Pfizer brought Terramycin to market: "No company would be transformed by broad-spectrum antibiotics as completely as Pfizer; and conversely, no company would change the marketing of antibiotics as much as Pfizer with its Terramycin, introduced in March 1950."[7] In addition to broad-spectrum antibiotics, a significant change in Pfizer's promotional techniques (more representatives and a greater advertising presence in trade journals, including the *Journal of the American Medical Association*) had a monumental effect. Between 1950 and 1956, antibiotic consumption in the United States increased from 139.8 to 645.2 tons, with antibiotics becoming the most commonly prescribed class of drugs.

However, this substantial increase in volume was not properly regulated. Although the Food, Drug and Cosmetic Act, promulgated by the FDA in 1938 in response to the tragedy of Elixir Sulfanilamide—the administration of which had caused the death of 107 people, including many children—was now supposed to guarantee the safety of available treatments, there were no means, other than via the opinion of the American Medical Association (AMA), of attesting to the efficacy of the drugs put on the market. In fact, the AMA began to forge closer and closer ties with the pharmaceutical industry, culminating in a growing number of accommodating arrangements, particularly in the 1950s, concerning the development of fixed-dose combinations of antibiotics, which were held up as a universal panacea for multiple infections, supposedly mitigating the risks of side effects by reducing the dose of each antibiotic administered, and reducing the risks of the emergence of resistance—problems that the medical profession was beginning to face. These combinations are based on the idea of synergy between antibiotics, which in reality is not guaranteed and is considered to pose significant problems by some specialists.

This is the context—marked on one hand by a significant increase in the volume and types of antibiotics produced and on the other by a regulatory void that allowed pharmaceutical companies to market more or less what they wanted—in which the call for controlled clinical trials in the late 1950s should be understood.[8] Senator Estes Kefauver's hearings on the pharmaceutical industry, beginning in late 1959 and culminating in the Kefauver–Harris Amendments of 1962, explicitly mandating proof of a drug's efficacy prior to FDA approval, marked a watershed event in twentieth-century therapeutics— an event inextricably linked to the development of antibiotics. From this date onward, no drug, antibiotic or otherwise, could be granted a marketing authorization without first being proven safe and effective. However, this new regulatory framework had no impact on prescriptions of antibiotics for therapeutic purposes, which continued to rise throughout the 1970s and up to the present day, though with disparities according to country and era.

The use of antibiotics was not limited to human health. As early as 1943, penicillin was introduced on farms in England and then in Denmark, where it was used to treat bovine mastitis and thereby maintain the quality of milk, which was particularly important in wartime. In the United States, mass medication of livestock is being developed in part thanks to the rapid production of new antibiotics; this reduces the risk of disease in increasingly dense populations and, at the same time, the costly labor required to care for sick animals.[9] However, the human and animal health historian Abigail Woods points out that antibiotics were not the only means of tackling epidemics in animal populations. In pig farming in particular, other methods, based on the development of an understanding of diseases as complex phenomena based on multiple relationships between animals and their environments, preceded antibiotics and continued to develop after their introduction into farming practices. These methods have contributed to the incredible productivity and performance gains of industrial agriculture.[10]

However, the uses of antibiotics have not remained solely therapeutic or prophylactic. At the end of the 1940s, the addition of antibiotics to poultry feed was authorized to prevent the occurrence of epidemics. As Hannah Landecker points out in a paper examining the history of feed production for farm animals, this feed already contained additives including vitamins, trace metals such as manganese and copper, growth promoters such as arsenic-based drugs, urea, and certain amino acids, as well as magnesium oxide or butyric acid to enhance appetite. The production of supplemented foods was the subject of intense research and constant toing and froing between academia and industry. It was in the context of extreme attention to the role of feed in increasing livestock productivity that researchers at Lederle first noticed and then demonstrated a significant effect of low-dose aureomycin on livestock growth.[11] Once again, the use of knowledge, correlations, and causalities proved never to be innocent. The historian Delphine Berdah has shown that, above all, this discovery, which remained quite controversial in the contemporary literature, created a new market for antibiotics and led to a reduction in the production costs of antibiotics for human use (because from then on, pharmaceutical production waste could be recycled).[12] Nevertheless, it triggered a massive rise in the consumption of antibiotics over the following years.

From growth and therapy to prophylaxis, antibiotics thus seemed to offer an effective and easy solution to a variety of problems. Enticed by the prospect of never-ending gains, manufacturers produced new antibiotics, forged new relationships, and developed other uses, notably to treat certain bacterial infections in plants. Antibiotics would soon also be used in food preservation and conservation.

As the historian Claas Kirchhelle mentions, although disparities had emerged in Europe (partly owing to differing national contexts, product availability, and the role of veterinarians), farmers were soon won over by the drop in the price of antibiotics and the promise of faster livestock growth with antibiotic supplementation.[13] By the end of the 1950s, up to 50 percent of British pigs were fed with such feeds, streptomycin-based sprays were developed to treat bacterial infections in fruit trees, and oxytetracycline was being used in Norway to preserve whale meat. In Japan, antibiotics also found their way into livestock and fish farms, followed by rice fields from the early 1960s onward. Between 1951 and 1970, there was an eleven-fold increase in the total use of antibiotics in the United States, from 690 to 7,670 tons, accompanied by a thirty-fold increase in the volume of nontherapeutic antibiotics (used, for example, in animal feed and for food preservation), rising from 110 to 3,310 tons. In France, around 30 tons of antibiotics were added to animal feed in 1964, while in Great Britain, around 41 percent of the total volume of antibiotics—168 tons—produced in 1967 was consumed by farm animals, half that in the form of feed additives. In the 1970s, antibiotics were used in salmon and other types of fish farms for tackling bacterial infections.

In addition, the United States was exporting its industrial agricultural model and the necessary production infrastructure to Africa, South America, and Southeast Asia, where governments wanted to modernize their agricultural practices, to the delight of US pharmaceutical companies and supplemental food producers, soon to be joined by other Western companies. A spectacular expansion in intensive livestock production was then witnessed as illustrated by the following examples. Between 1968 and 1998, there was a twenty-fold increase in Brazilian chicken production. In China, the introduction of free-market economic policies from the 1980s onward led to the establishment of American and Thai companies and the growth of intensive poultry farms in the country. In the first decade of the twenty-first century, an unprecedented increase was recorded in the pork sector, with China producing four times as many pigs as the United States in 2008. The figures quoted by Kirchhelle are edifying: According to a 1997 report, 750 to 1,000 tons of chlortetracycline and 5,000 to 7,000 tons of oxytetracycline were used annually to feed around 500 million pigs, 36 million cattle, and 70 billion poultry. In 2010, China became the world's largest consumer of agricultural antibiotics (accounting for approximately 23 percent of global use), but, as we shall see, these transformations have also entailed modifications to the organisms themselves through the selection and transformation of farm animals to adapt them to ever-more massive modes of production.[14]

While antibiotics enable the treatment of infectious diseases that might have been fatal only a few decades ago, their use has since spread to all sectors, as shown by the detailed studies conducted by numerous social science researchers.[15] As the anthropologists Laurie Denyer Willis and Clare Chandler have shown, in low- and middle-income countries, antibiotics are a "quick fix" for hygiene problems that, contrary to what is often claimed, cannot simply be explained by individual behavior.[16] Rather, they are linked to the inequalities that structure these societies, and appear to be the consequence of a series of neoliberal reforms, structural adjustments and policies to marginalize the poorest and most vulnerable populations.[17] Antibiotics are therefore used as quick fixes in landscapes marked by shortages, uncertainties, and inequalities, where they also solve productivity problems by getting humans back to work faster when they are sick. This practice is explicitly mentioned in the World Health Organization's 2015 *Global Action Plan on Antimicrobial Resistance*, which equates human losses resulting from antibiotic resistance to productivity losses that can be quantified and translated into euro equivalents, thereby reducing individuals to mere tools of production.[18] Consequently, "antibiotics are deployed to paper over long-term structural issues that undermine care provision, drive increased productivity and correct for hygiene issues caused by entrenched inequality."[19] This observation, far from being limited to low- and middle-income countries, can be extended to the entire world.

SCALABILITY AND PLANTATION

While antibiotics and their many uses can be analyzed in public health terms, they should also be situated in the longer history of the exploitation of nature and living beings, in which they play an active role. This leads me to borrow the "term *plantation* in its broadest sense, [which evokes] simplified ecologies designed to create assets for future investments," from the anthropologist Anna Tsing.[20]

The plantation system is based on the orderly, systematic, and standardized exploitation of the labor and reproduction of the species grown on a plantation and of the humans who work there. This exploitation requires standardization and therefore the commensurability and quantification of various living entities, both human and nonhuman. Just as the standardization of model organisms in biology laboratories requires a reduction in the parameters to be considered and a simplification of the living environment, and just as the production of commensurable data in clinical trials results from the neutralization of various idiosyncrasies, such objectification of living beings *for the purpose of producing them* can be effective only when they are removed from their usual living environment so as to reduce the potential of their interaction with other species. In this sense, we cannot conceptualize the idea of *plantation* without simultaneously contemplating the *eradication* of species and relationships considered harmful to the project. Historically, plantations have shown to quickly become single-crop farms, worked by human beings who have been displaced, uprooted from their lands and reduced to slavery, or exploited for extremely low wages in an integrated system designed to extract everything that each being—human and nonhuman—is able to provide and requiring drastic means of control and coercion. Plantations have sometimes developed in locations far removed from the original living environments of the species being cultivated. Each extension, each displacement, requires the reproduction of the conditions needed for the production of the organisms in question, whether human or nonhuman. However, these systems have proven efficient and effective, if considered in light of their objectives: to produce more and more at lower and lower costs. For this reason, they have become true role models.

Sidney Mintz, for example, has shown how Caribbean sugar plantations formed a proto-industrial labor model that shaped nascent industrialization in Europe.[21] Similarly, for Raj Patel and Jason Moore, the cane plantation was "a forerunner not only of today's industrial agriculture but of today's modern factory.

These early modern sugar plantations not only were highly mechanical, with large, fuel-intensive boilers and heavy-duty rolling mills to extract cane juice from stalks, but also served as powerful drivers of "simplification": of the work process, as workers (slaves) were given simplified tasks, and of the land itself, which was reduced to a cane monoculture. Just as auto-workers on the line assemble simplified, interchangeable parts and fast food workers manufacture standardized burgers."[22]

While the plantation model is still based on these simpli-fied ecologies, its scale has changed. Tsing borrows the notion of scalability from computer science to evoke "the ability of a project to change scales smoothly, without any change in proj-ect frames."[23] Using terms such as *scalability*, *scaling up*, or any other that reflects a change of scale in the relationship that some humans maintain with living beings is not surprising given the history of antibiotics. The intensification of breeding and the homogenization of livestock and plant varieties through genetic selection or cloning have been undertaken to increase the profitability of each living being that is cultivated or bred (e.g., accelerated growth, selection of plants giving larger grains), as well as the profitability of their processing, first by humans and then by machines. These changes in farming methods and in the biology and physiology of farmed populations make those populations more vulnerable to outbreaks of pathogenic micro-organisms. Drastically increasing the concentrations of animals in an environment leads to an inevitable increase in the trans-mission of pathogenic germs (there is no "social distancing" on farms). This risk is further increased by the low genetic and phe-notypic diversity of the individuals forming the herd or the crop, meaning that all are likely to have a similar or identical response to encounters with the same pathogen. This means that if a pig (or a chicken, or a corn plant) is vulnerable to a bacteria (or a virus, or a fungus), the vast majority of the herd (or the culture) will also be vulnerable. If this pig also shares a confined space with other pigs, then the transmission will be even faster. Antibiotics, used as a preventive or curative measure, can drastically reduce the risk of infection, thereby increasing the size, productivity, and profitability of farms. Moreover, antibiotics enable us to keep pushing the limits of the standardization of living beings. Such a process is chillingly depicted by Alex Blanchette in *Porkopolis*, which provides a detailed description of hog production in the factory farms of a Midwestern city. Blanchette recounts how new forms of division of labor and new competencies are being developed that even transform workers' bodies. He also explains that humans are single-mindedly devoted to the reproduction of a particular species of ultra-standardized hog (just as scientists in laboratories are single-mindedly devoted to the production and reproduction of model organisms). In this configuration, even the managers of the firm studied by Blanchette admit that the sole means of further increasing production rates and productivity are to create an ultimate standardization capable of producing perfectly morphologically homogeneous hogs that could be butchered by machines rather than humans.[24]

The history of the industrialization of antibiotics teaches us that these chemical molecules have been and remain a precondition for the scalability of the reproduction and exploitation of living beings, be they humans, animals, or plants. Today's plantations—fish farms, shrimp farms in Bangladesh, citrus plantations in California, and, in a perfect encapsulation of Tsing's plantations and Patel and Moore's factories, factory farms in China and the American Midwest—would be unthinkable without antibiotics.[25] Such degrees of standardization and intensification could not be envisaged without the precious chemical molecules produced and administered by the ton to

virtually clonal populations (and even to totally clonal populations where plants are concerned), whose growth they accelerate
and which they protect from certain pathogens. Nor would they
be feasible without increasing the productivity of the labor force
responsible for extracting the value from these living beings that
such molecules make possible by keeping them in reasonable
health or getting them back to work faster after an infection.
To the point that we can talk about a successful pairing between
a therapeutic solution—antibiotics (and the pharmaceutical
industry in general)—and the development of forms of capitalism based, since the Second World War, on mass production and
consumption, as well as the standardization of procedures, products and devices, and even tastes. Antibiotics, massively produced and consumed, pave the way for the intensification and
massification of production, which in turn requires more and
more antibiotics, both quantitatively and qualitatively.

THE PLANTATIONOCENE AND
THE CAPITALOCENE

For several years, many scientific experts have been debating the
need to characterize a change of geological era that can account
for the impact of human activities on the documented changes
in the climate and in the earth as a whole—changes that can
be witnessed at all levels, even in sedimentary deposits. These
debates revolve around the name of this new era and the date
on which it should be said to have begun.[26] Choosing the right
name is important. *Anthropocene*, as many have pointed out, ultimately places the burden for these changes on all human beings:
on an undifferentiated *anthropos*, when it is widely documented
that the contributions of various populations, and even the

various social classes in these populations, are far from equal. Now let us consider the period. Some people believe that the Anthropocene should coincide with the beginnings of domestication, that is, the Neolithic period. This anthropocentric conception is weak from an analytical standpoint as it disregards a dimension central to this book, namely the questions of evolution (of, for example, species, societies, lifestyles, knowledge, and technologies) and, more generally, of pluribiosis. Others link the beginnings of the Anthropocene to industrialization and the exploitation of fossil fuels, considering the industrial revolutions of the nineteenth century to be the tipping point. However, I consider a primary focus on energy sources and the technologies enabling their exploitation to be a limitation, one representative of the difficulty in conceptualizing the intricate links of interdependency between living beings and the all-encompassing nature of the current ecological crisis, between human and nonhuman living beings, as well as between biotic and abiotic factors.[27]

To account for the profound changes in relationships among living beings, which are apparent in the plantation model, Anna Tsing, Donna Haraway, and Scott Gilbert prefer the much more precise and *relational* term *Plantationocene* to the undifferentiated *Anthropocene*.[28] Adopting *Plantationocene* means describing modes of relationships and defining a period according to the types of interspecies relationships that predominate in it. To put it another way, it means defining a period according to the type of ecology that is preferentially developed in it, even if that ecology is based on a gross oversimplification of relationships between living beings. The Plantationocene corresponds to the reign of anti-pluribiosis, of a conception of living beings as fixed entities; it is an era in which living beings become all the more controllable because they are deprived of the interrelationships they need. This conception is close to that of Jason Moore, who prefers the term *Capitalocene* for its emphasis on unequal exchanges, appropriation, exploitation, and accumulation and to stress that capitalism is not only an economic system but also a specific set of *relations* between humans and the world (biotic and abiotic).[29] The plantation was one of the first incarnations of such a set of relations, and it benefited from successive evolutions, driven by both industrial technologies and governance projects, notably imperial ones.[30] The Capitalocene embodies the ever-increasing scalability of the plantation model.

Authors like Tsing, Haraway, Gilbert, and Moore also agree on the historical demarcation of the Plantationocene or Capitalocene epoch. The origin of this type of relationship, underlying both the plantation and scalability, can be situated in the sixteenth century, marking the transition from feudalism to capitalism. This period, characterized by numerous transformations, is intrinsically marked by the consequences of Cartesian conceptions, which authorize the separation of nature and culture and permit humans to "behave as the masters and owners of Nature." Moore refers to the act of separating nature and society as the "externalization of nature": including and excluding, recognizing or ignoring according to context, and breaking the bonds of dependence.[31] Some beings are placed alongside humans, whereas others are situated in a nature perceived as external, which makes them and the work they perform appropriable and exploitable. This category encompasses not only nonhuman living beings—"natural resources" in today's parlance—but also most of humanity, whether colonized populations or women. Situating living beings and resources in an "external" nature to better appropriate them was the motivation for the enclosure movement that began in sixteenth-century England, a motivation that also explains the "enclosure" of women and their

Earth system trends

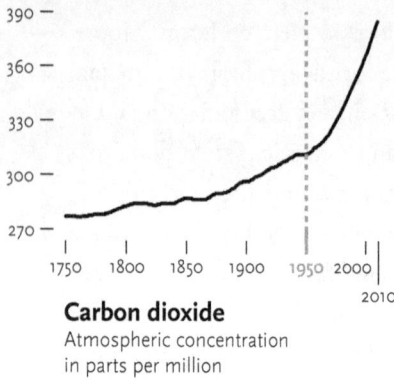

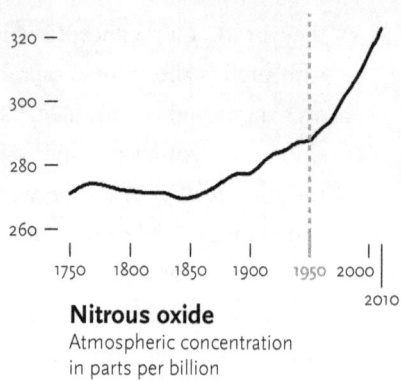

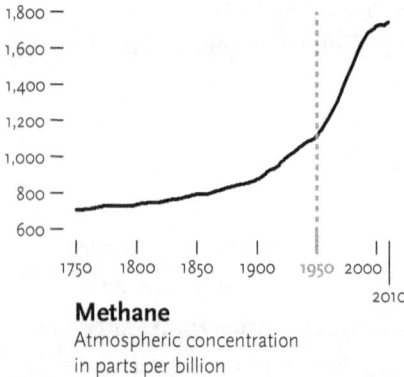

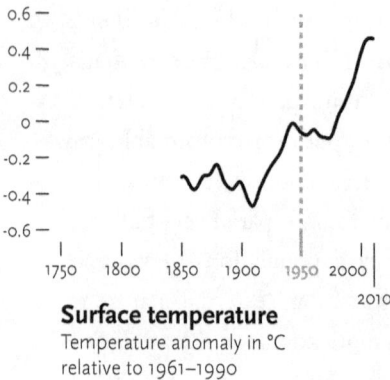

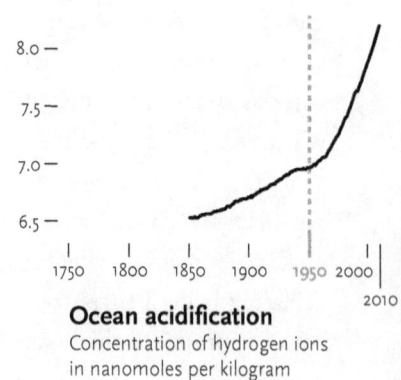

Carbon dioxide
Atmospheric concentration
in parts per million

Nitrous oxide
Atmospheric concentration
in parts per billion

Methane
Atmospheric concentration
in parts per billion

Stratospheric ozone
Percent loss

Surface temperature
Temperature anomaly in °C
relative to 1961–1990

Ocean acidification
Concentration of hydrogen ions
in nanomoles per kilogram

FIGURE 7.2 Earth system evolution

Source: Adapted from Will Steffen et al., "The Anthropocene: Are Humans Now
Overwhelming the Great Forces of Nature?" *Ambio* 36, no. 8 (2007): 614–21.

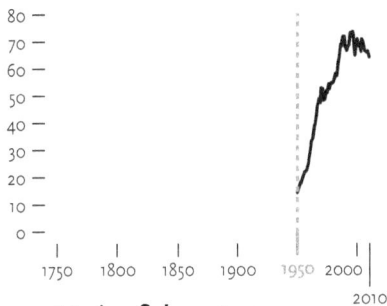

Marine fish capture
in millions of tons

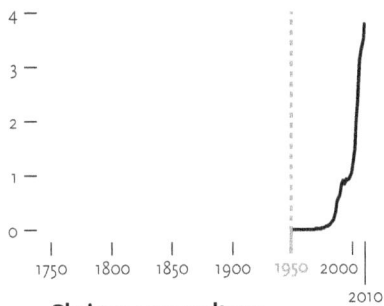

Shrimp aquaculture
in millions of tons

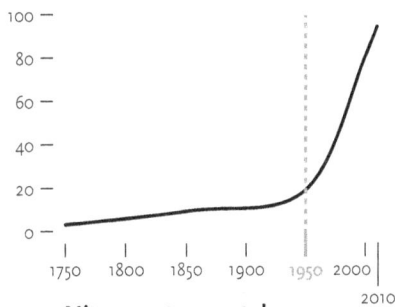

Nitrogen to coastal zone
Human-induced flux
in millions of tons per year

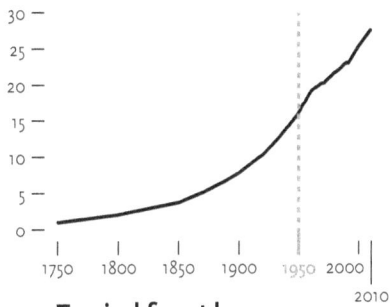

Tropical forest loss
Percent loss (since 1700)

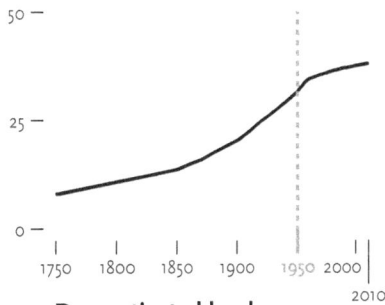

Domesticated land
Percent of total land area

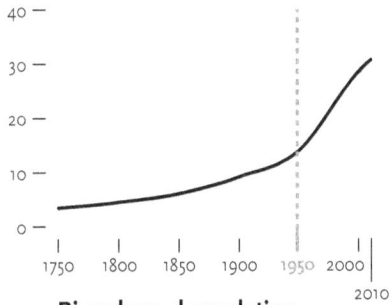

Biosphere degradation
Percent decrease in terrestrial
mean species abundance

Socio-economic trends

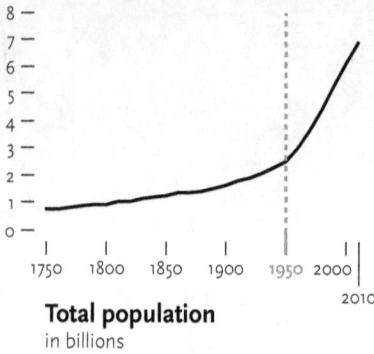

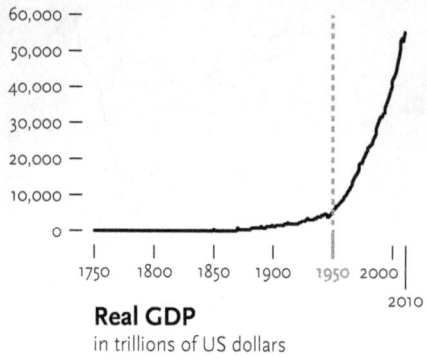

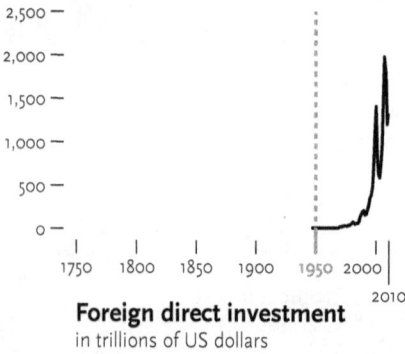

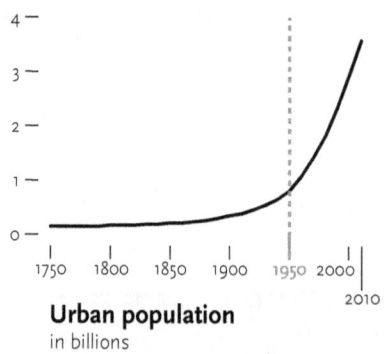

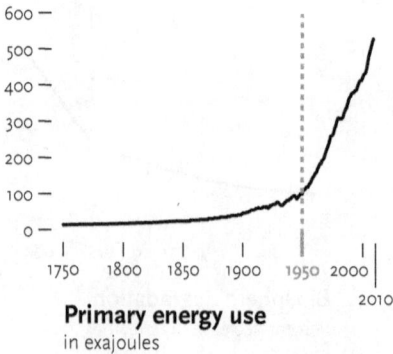

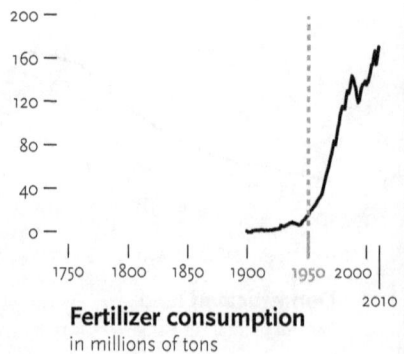

Total population
in billions

Real GDP
in trillions of US dollars

Foreign direct investment
in trillions of US dollars

Urban population
in billions

Primary energy use
in exajoules

Fertilizer consumption
in millions of tons

FIGURE 7.3 Socioeconomic development

Source: Adapted from Will Steffen et al., "The Anthropocene: Are Humans Now Overwhelming the Great Forces of Nature?" *Ambio* 36, no. 8 (2007): 614–21.

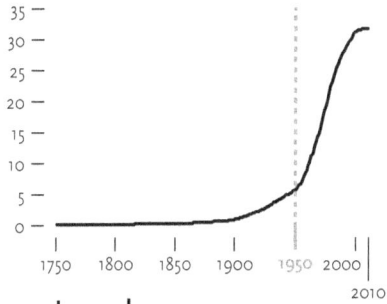

Large dams
in thousands

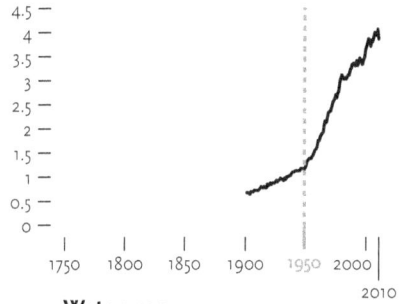

Water use
in thousands of km³

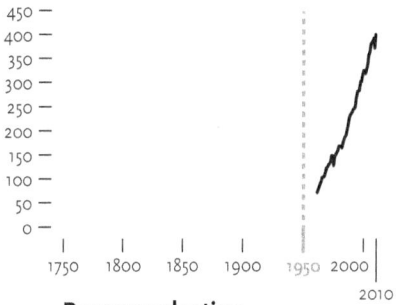

Paper production
in millions of tons

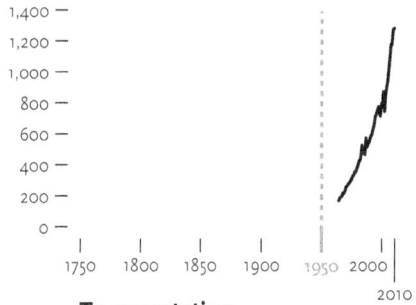

Transportation
in millions of motor vehicles

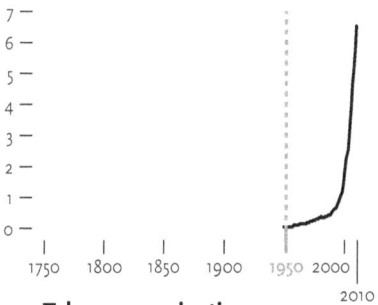

Telecommunications
in billions of phone subscriptions

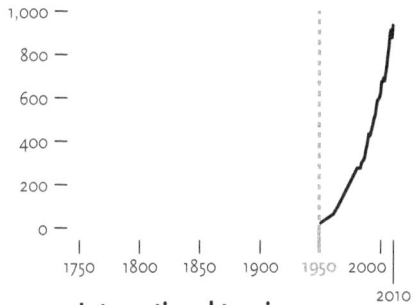

International tourism
in millions of arrivals

confinement to the domestic space, in contrast to the Middle Ages, when they participated in all trades.[32] In addition, externalization stems from the exploitation and invisibilization of the work carried out by living beings excluded from a common humanity: slaves on the plantations, women who do most of the work associated with the *reproduction* of the labor force, as the Marxist feminists have demonstrated, and even nature itself.[33] The invisibilization of the work associated with *reproduction* would enable the production of *cheap* forms of nature.

In *A History of the World in Seven Cheap Things*, Raj Patel and Jason Moore define *cheapening* as "a strategy, a practice, a violence that mobilizes all kinds of work—human and animal, botanical and geological—with as little compensation as possible. We use *cheap* to talk about the process through which capitalism transmutes these undenominated relationships of life making into circuits of production and consumption, in which these relations come to have as low a price as possible."[34] Colonizing new territories to exploit their inhabitants, soils, and subsoils is one of these processes.[35] Capturing and transforming arable land is another, as is increasing and concentrating livestock to maximize productivity in a process of ever-increasing standardization and of constantly breaking links of interdependence to produce more and more, with less labor in less space.

In this centuries-long history, the introduction of antibiotics coincides with what Will Steffen, Paul Crutzen, and John McNeill call the "great acceleration." This period, which began at the end of the Second World War, is characterized by an exponential increase in a certain number of markers.[36] These markers reflect both the exacerbation of the epoch's characteristic phenomena and the impacts of pernicious ecologies: a massive increase in the presence of certain compounds such as carbon dioxide, nitrous oxide, methane, and stratospheric nitrogen in the atmosphere; steady and rapid increases in the earth's temperature, the acidification of the oceans, and the percentage of domesticated land, as well as in the number of species going extinct and the hectares of primary forests being ravaged; and the growth in the consumption of energy, fertilizers, water, livestock (both terrestrial and marine), and autoimmune diseases (figure 7.2 and 7.3). Antibiotics do not feature in the various diagrams on the great acceleration, yet they, too, have increased exponentially.[37] While the roles of antibiotics in the profound transformations of capitalism in the second half of the twentieth century would require a detailed analysis, as their roles in the various diagrams, the constant increase in human population would be difficult to understand without also understanding the drastic decrease in mortality (especially in infants) we have experienced thanks to antibiotics. It would be difficult to understand the increases in atmospheric methane levels and in the consumption of water and fertilizers without relating them to the intensification of agriculture largely permitted by these precious chemical molecules, which also enabled increases in productive *and* reproductive forces. The "quick fixes" described by Denyer Willis and Chandler in low- and middle-income countries are just one contribution to the production of "cheap" forms of nature. At the very least, antibiotics make it possible to break new ground in the process of "cheapening": cheaper and cheaper healthcare, ever-cheaper food, and exceedingly cheap work.

Antibiotics are wonder drugs, the "magic bullet" of medicine; at least, they used to be. But in the second half of the twentieth century, they became much more than that. Seized upon by capitalist systems of relationships, mainly for their capacity to massively eradicate certain microbes, they have also participated in plantation economies that rely on the standardization, reduction, and appropriation of living beings—both human and nonhuman—by enabling the development of new forms of exploitation previously imagined as occurring only in dystopias and of which the factory farms of *Porkopolis* are but one example among many others.

Anna Tsing tells us that plantations "also support new ecologies of proliferation: the unmanageable spread of life from plantations, in the form of disease and pollution." This is certainly where we see the most brutal manifestation of antibiotics as pharmakon, both remedies and poisons. These molecules have been and continue to be used in efforts to put an end to this unmanageable spread: to treat or prevent the bacterial epidemics that inevitably spread within and outside plantations. However, such antibiotics have not remained confined to plantations; rather, they have themselves led to other forms of proliferation and propagation.

Antibiotics become visible as infrastructures when problems arise, revealing a number of elements that had been taken for granted or not questioned: *Diseases once made benign thanks to antibiotics are now becoming lethal again.*

8

RECALCITRANCE AND FERALITY

By chemical infrastructure I mean the spatial and temporal distribution of industrially produced chemicals as they are produced, consumed, become mobile in the atmosphere, settle into landscapes, travel in waterways, leach from commodities, are regulated (or not) by states, monitored by experts, engineered by industries, absorbed by bodies, metabolized physiologically, bioaccumulate in food changes, break down over time, or persist. Chemical infrastructures are regulated and ignored, studied and yet filled with uncertainties. . . . Chemical infrastructures, importantly, are spatially and temporally extensive. They are translocal, connecting moments of production and consumption, moving across borders and traversing scales of life. They are temporally distributed, as some chemicals break down quickly, and others persist and are present for the longue durée, some causing immediate responses in organisms, others only provoking effects that take generations to see, working a slow injury on ecologies and organisms, or even planetary atmospheres.

A few years before Clare Chandler proposed considering antibiotics as infrastructures, the anthropologist Michelle Murphy, in an edifying paper on the petrochemical history of the St. Clair River and "Chemical Valley" in Canada (where 40 percent of Canadian petrochemical processing takes place), was already talking about chemical infrastructures as "constructed ecologies" that go far beyond the framework in which the humans who produced them intended to confine them. This is because chemical substances have their own life, their own agency, and their own particularities. Some degrade rapidly, whereas others migrate not only in space but also over time. Chemical molecules "leak"; far from being stable objects, they are things, as the anthropologist Alex Nading reminds us.[1] They leak and escape, belying the fictions of chemical infrastructures remaining "confined" in space and time and dismantling "the regulatory fantasy that chemical infrastructure is a matter of contained security."[2] They leak but leave traces that act as invitations to track down and identify the origins of them in their infrastructures. However, how they leak, and therefore the responses that could be proposed, are highly dependent on the agency of the chemical molecules in question. The demonstration that I shall present in this chapter is therefore only valid in the case of antibiotics, although parallels with other molecules (such as chlordecone) can be drawn and commented on.

The phenomena of chemical leakage can also be understood from the perspective of what Anna Tsing calls "ferality." *Ferality* refers to the act of returning to a wild state after being domesticated. Feral entities are those that escape from built ecologies, plantations, and infrastructures and that develop and spread beyond human control.[3] Ferality is also another way of talking about the recalcitrance of microbes, about their capacity to do something other than what is expected of them by the humans who manipulate them. They are an additional manifestation of pluribiosis and a reminder that humans should never forget that the categories they create, whatever they may be, are constructed, that the stable and fixed outlines assigned by the life sciences to the entities with which they work, while particularly useful for conceptualizing and acting on the world, remain abstractions. Microorganisms have enormous and unimaginable potentialities: Bacteriophage viruses represent the largest reservoir of genes in the world, which also makes them the largest reservoir of all kinds of evolutionary surprises.

A BRIEF HISTORY OF
ANTIBIOTIC RESISTANCE

The acquisition of bacterial resistance to antimicrobial molecules is a known phenomenon that was documented early.[4] Through processes of mutation or gene exchange, which I shall discuss in more detail later, some bacteria acquire the ability to resist the effects of a given antimicrobial molecule. Antibiotics employ numerous mechanisms of action, but bacteria can employ just as many ways to counteract their effects, and since the discovery of penicillin, significant clinical resistance has rapidly followed the rollout of each new antibiotic. This was the case for methicillin, to which the first forms of resistance were documented as early as 1961, one year after its addition to pharmacopoeias.

Scientists and clinicians were quick to issue warnings about resistance phenomena. In 1945 in his Nobel Prize acceptance speech, Alexander Fleming warned of the risks associated with the misuse of antibiotics:

It is not difficult to make microbes resistant to penicillin in the laboratory by exposing them to concentrations not sufficient to kill them, and the same has occasionally happened in the body. The time may come when penicillin can be bought by anyone in the shops. Then there is the danger that the ignorant man may easily underdose himself and, by exposing his microbes to non-lethal quantities of the drug, make them resistant. Here is a hypothetical illustration. Mr. X has a sore throat. He buys some penicillin and gives himself not enough to kill the streptococci but enough to educate them to resist penicillin. He then infects his wife. Mrs. X develops pneumonia and is treated with penicillin. As the streptococci are now resistant to penicillin, the treatment fails. Mrs. X dies. Who is primarily responsible for Mrs. X's death? Why Mr. X, whose negligent use of the penicillin changed the nature of the microbe. Moral: If you use penicillin, use enough.[5]

Fleming's discussion of the misuse of chemical molecules focused on the risks associated with underdosing ("If you use penicillin, use enough"), an action he attributed to individuals who were "ignorant" or "negligent" but "responsible" for their actions. He presented resistance as the result of punctual interactions between humans and bacteria in which the bacteria can "educate" themselves or "change their nature," if not eradicated.

Indeed, bacterial resistance in the midtwentieth century was mainly interpreted as a process of adaptation through point mutations and selection. Broadly speaking, this means that if you put a population of bacteria in a given environment, they will multiply, but with each successive generation, mutations will appear because of imperfections in the replication processes of genetic material. The vast majority of mutations will not produce significant differences, but some may confer a better adaptation to the environment, a selective advantage. If you add an antibiotic to a medium, almost all the bacteria sensitive to that antibiotic will die. A natural resistance or a mutation in a bacterium is sufficient for it to survive and multiply, transmitting this ability vertically (i.e., to its descendants) and thus allowing

Bedaqui- ◉
Fidaxomicin ◉
Glycylcycline (Tigecycline) ◉∅
Daptomycin ◉
Oxazolidinone ◉∅
∅——◉ Ciprofloxacin
Carbapenem ◉——∅
◭ Quinolones
Ampicillin ∅————◉
◭ Fusidic Acid
Streptogramins ◭
Cephalosporin ∅————◉
◉∅ Methicillin (Beta-lactams)
Colistin (Polymyxin) ∅———————————————◉
∅—◉ Rifampin
Vancomycin (Glycopeptides) ∅——————————◉
Erythromycin ∅——————————◉
Aminoglyco- ◉
◭ Chloramphenicol
∅——◉ Tetracy-
Penicillin (Beta-lactams) ◉∅
Sulfonamides ∅——∅◉·······◉
1905
∅————◉ Arsphenamine (Salvarsan)
2015

1910 1920 1930 1940 1950 1960 1970 1980 1990 2000 2010

∅ : First clinical uses of the antibiotic
◉ : Identification of resistance in a bacterium
◉ : Clinical use and identification of resistance in the same year

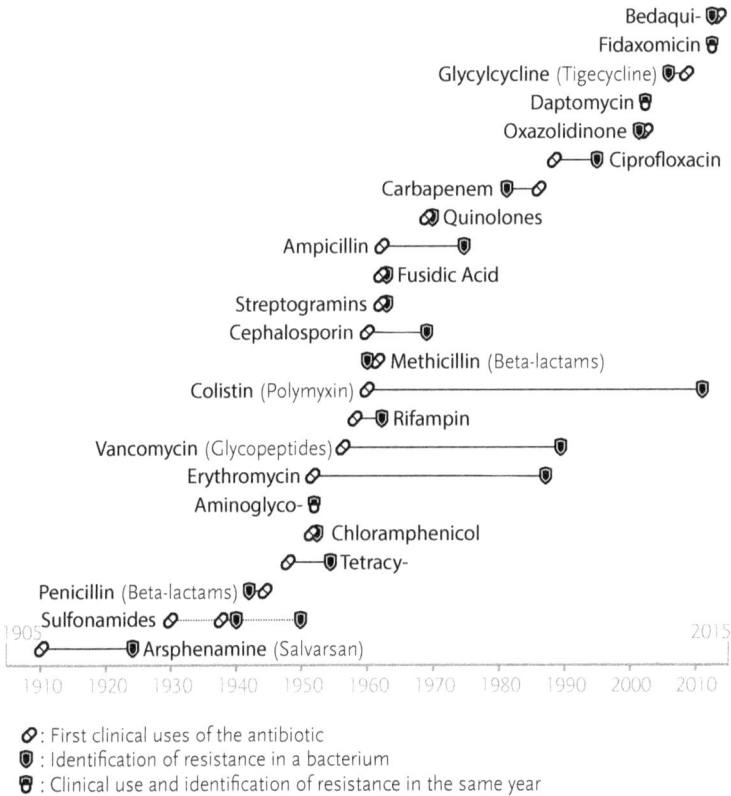

FIGURE 8.1 Emergence of antibiotic resistance, 1905–2015

Source: Adapted from Fabrizio Spagnolo et al., "Why Do Antibiotics Exist?"
mBio 12, no. 6 (2021): e01966–21.

for the repopulation of the environment. Hence Fleming's statement that antibiotic doses that are too low allow more time and therefore more opportunities for certain individuals in a bacterial population to evolve and thus to develop resistance. But until the 1960s, pharmaceutical companies' optimism and confidence in their ability to develop new molecules prevailed.[6]

However, in the 1950s, Esther Lederberg and her husband, Joshua, had already demonstrated a phenomenon described in chapter 2: transduction, the process by which bacteriophages, while passing from one bacterium to another, facilitate exchanges of bacterial DNA. For this research, Joshua (but not Esther) was awarded the Nobel Prize in Physiology or Medicine in 1958, along with George Beadle and Edward Tatum. The Lederbergs were also working on the biology of plasmids: circular fragments of DNA that can contain several dozen genes and are sufficiently stable to resist being degraded too quickly outside a cell. Plasmids can be exchanged between bacteria during *conjugation* (figure 8.2), but they can also be "ingested" directly by a nearby bacterium when released into the environment—after bacterial lysis, for example. At that time, these microbial properties were mainly used as tools in laboratories in the development of molecular biology (as discussed in chapter 2). However, they were also mobilized by a team of Japanese scientists confronted with a series of dysentery epidemics in the late 1950s. Infected individuals carried strains of the bacterial species *Shigella* and strains of *Escherichia coli* that were peculiar for their resistance to multiple antibiotics and their ability to transmit this multiple resistance to other bacteria. Exchanges of antibiotic-resistance genes could therefore take place in a single block between different bacterial species. In 1959, the team, led by Tsutomu Watanabe, succeeded in isolating an extrachromosomal DNA molecule (i.e., a molecule not carried by one of the bacterial chromosomes and therefore not inserted into those chromosomes), which they called an "R factor" ("R" for "resistance"). R factors have since become known as plasmids.[7]

Also in 1959, the British geneticist Naomi Datta was assigned to investigate an outbreak of *Salmonella typhimurium* at Hammersmith Hospital in London. Two years later, after testing the

Conjugation mediated by the F factor

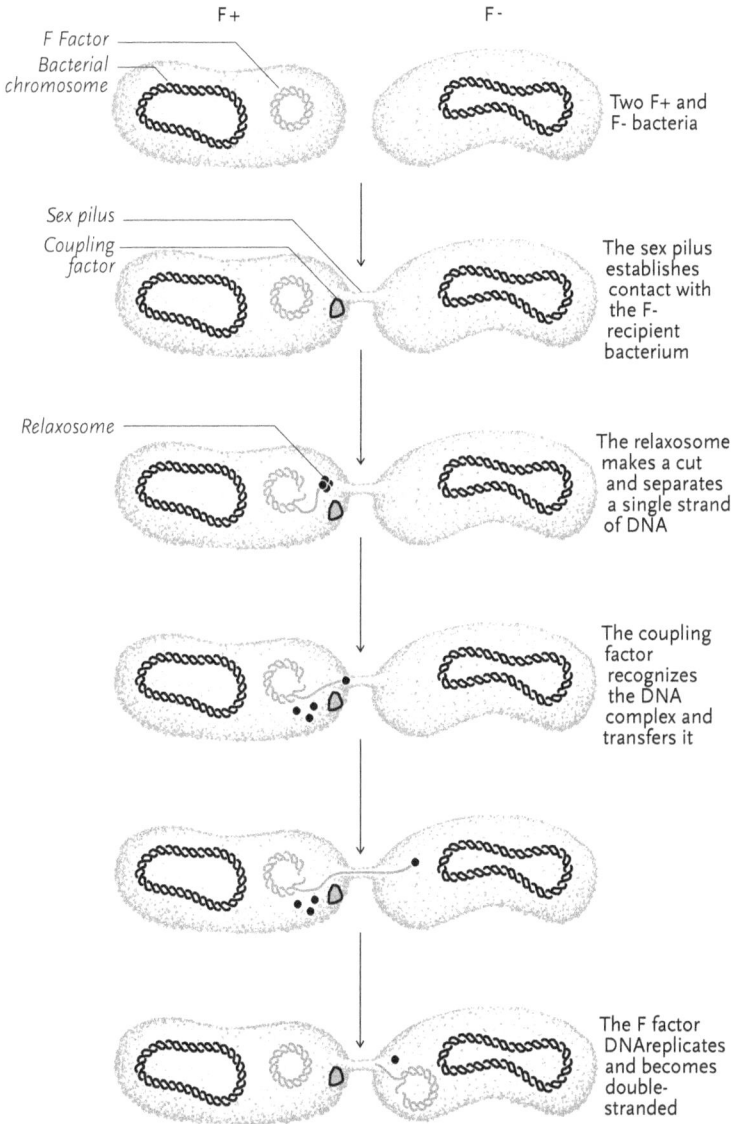

F+   F-

F Factor
Bacterial chromosome

Two F+ and F- bacteria

Sex pilus
Coupling factor

The sex pilus establishes contact with the F- recipient bacterium

Relaxosome

The relaxosome makes a cut and separates a single strand of DNA

The coupling factor recognizes the DNA complex and transfers it

The F factor DNA replicates and becomes double-stranded

FIGURE 8.2 Conjugation

antibiotic resistance of bacteria in stool samples from more than three hundred patients, her team also concluded that an R factor was present. In 1963, Datta began collaborating with the bacteriologist Ephraim Saul Anderson and other scientists. Their work revealed the rapid emergence of multiple-antibiotic resistance in enterobacteria and that it was linked to a sharp rise in antibiotic use. However, they also showed that the transfer of multiple resistance, not only between bacterial species but also between their hosts, had already occurred on British farms, with resistant bacteria passing from cattle to farmers, veterinarians, and their families.[8] These studies led to a series of reports by Anderson challenging the existing understanding of resistance risks. Anderson advocated restricting the use of medically relevant antibiotics to humans to prevent the emergence of multiple forms of resistance in particularly dangerous pathogenic bacteria.

It was not until 1968 that, under pressure from public opinion, the British authorities commissioned a study on antibiotics, leading to the recommendation of a series of reforms in 1969, starting with a ban on the use of therapeutic antibiotics such as penicillin and tetracyclines as growth promoters in livestock. However, these recommendations had a limited impact: Although Great Britain, Switzerland, and some European Economic Community member states adopted them, veterinarians could override the bans simply by issuing more therapeutic prescriptions. In the United States, a 1966 report highlighting the uncertainty concerning the link between antibiotic use in agriculture and the development of antibiotic resistance in humans prevented the implementation of the same type of ban.[9]

Although the World Health Organization began convening working groups on antibiotic resistance in the second half of the 1970s, antibiotic resistance did not become a global concern until 1981, thanks to the efforts of the researcher and clinician Stuart Levy. That year, Levy organized a conference in the Dominican

Republic titled "Molecular Biology, Pathogenicity and Ecology of Bacterial Plasmids," at which 147 scientists from twenty-seven countries signed a joint declaration on the misuse of antibiotics worldwide. Despite this declaration and the efforts of many scientists over the following years, the problem of bacterial antibiotic resistance remained largely ignored by many governments.[10] In the years since, antibiotic resistance has been linked to broader concerns about emerging infections, focusing increased attention and funding toward the issue. It has also been progressively integrated into the programmatic framework of the "One Health" initiative linking human, animal, and environmental health, inspired by the Wildlife Conservation Society and officially supported since 2008 by the World Health Organization, the World Organisation for Animal Health, and the Food and Agriculture Organization.[11]

According to Podolsky's analysis, however, it was not until the 2013 publication of *Infections and the Rise of Antimicrobial Resistance*, a report written by Dame Sally Davies, that a turning point was reached. In December 2013, the United States began to address the use of antibiotics in agriculture, imposing restrictions on the use of medically important antibiotics as growth promoters. This change is attributable to Davies's report, likely to the following statement, which has since been widely repeated in discourses and cited in the literature: "Antimicrobial resistance is a ticking time-bomb, not only for the UK but for the world. We need to work with everyone to ensure the apocalyptic scenario of widespread antimicrobial resistance does not become a reality. This threat is arguably as important as climate change for the world."[12]

Infrastructures leak like a sieve, and Davies presents resistance not as a current burden but rather as a future emergency, a disaster that can be averted if the right choices (to plug the holes) are made immediately.

LIVING IN FUTURES PRODUCED
BY PREVIOUS STATES OF
MICROBIOLOGICAL KNOWLEDGE

The present is not a transparent given. The choices that must be made to tackle antimicrobial resistance are far from obvious and result from both the current state of knowledge (i.e., the knowledge available and the use to which it is put) and the active production of ignorance (which is as applicable to global warming as it is to antibiotic resistance). However, these choices also depend on the effects of the use of this knowledge. While Davies presents widespread antimicrobial resistance as a future emergency that we must anticipate, the historian of science Hannah Landecker shows that, as far as antibiotics and antibiotic resistance are concerned, we live in futures produced by previous states of knowledge, in this case specifically microbiological knowledge.[13]

What is now called "horizontal" or "lateral" gene transmission—which occurs via the mechanisms of conjugation, transduction, and transformation—was long considered insignificant; however, this phenomenon is now recognized as considerably important. This shift has had innumerable consequences for the life sciences. Some evolutionary specialists have raised serious doubts about the metaphor of the tree of life as a way of conceptualizing and illustrating the evolution of species, preferring more rhizomatic graphical representations, which are better suited to accounting for horizontal gene exchanges.[14] Others highlight these exchanges as a means of challenging controversial notions such as those of species and, more generally, the various classification systems available to scientists, considered too rigid and fixist, from a completely new perspective.[15]

This change in the conception of microbiology was made possible by the increase in resistance phenomena observed in both hospitals and research laboratories. This increase is itself a direct consequence of the increase in the production and consumption of antibiotics during the second half of the twentieth century, which was initially driven by a joyful optimism (based on the preeminence of vertical transmission in evolutionary models) about the pharmaceutical industry's capacity to discover increasingly new molecules and then by the ever-greater integration of these molecules into human lifestyles. What has made this integration possible are the incessant disputes about the link between antibiotic use and the emergence of antibiotic resistance, which have blurred the lines between objectified knowledge and the production of ignorance and thereby altered both the capacities and the will to act.[16] However, as Landecker tells us, we live in futures produced by what we thought we knew. To complete the picture, we could add "and by what we did not know that we did not know." This message serves as a warning at a time when the international community is looking for solutions.

This warning is issued on two levels. The first is already widely stressed by all social science and life science researchers studying antibiotic resistance. Everyone makes the same somewhat desperate observation that the policies of reasoned antibiotic management (referred to as "stewardship" in international texts) have been designed primarily to change individual behaviors (those of patients and those of clinicians) by means of slogans or incentives. Many people in France will remember the "Les antibiotiques, c'est pas automatique !" campaign based on a phrase whose infantilizing nature permeated even its grammatical structure (its literal translation being "Antibiotics, it's not automatic!"). In line with Fleming's discourse, individual responsibility is still the order of the day, through plans to educate individuals with the aim of preventing them from "educating" their bacteria using terms reminiscent of those used by André in chapter 1. Meanwhile, the systemic causes of antibiotic resistance are only timidly underlined in reports from international organizations and seem to be politely ignored in global governance.[17]

The second level of warning is more complex and based on the pluribiotic capacities of living beings. The history of antibiotic resistance is a quantitative *and* qualitative problem. First, quantitative: Far from being a problem simply of underdosing in humans, it is a problem of underdosing that applies to the entire planet. Counterintuitively, one could say that the widespread use of antibiotics in the second half of the twentieth century led to an exposure that was massive but mainly in the form of permanent underdosing at the local level—at the scale of bacterial populations—thus giving bacteria all the latitude they needed to evolve, adapt, and exchange. This brings us to an intrinsically qualitative problem, residing in the very mode of discovery of these molecules: Because they have been produced industrially for decades, we tend to forget that they are produced by microorganisms themselves to mediate their relationships with the environment. Indeed, the molecules produced by microorganisms constitute signals, all of which contribute to complex and subtle ecosystems while influencing dynamic equilibria, and all of which participate in microgeohistories. At the beginning of the "antibiotic era," entire networks of scientists, mycologists, bacteriologists, and soil scientists began exploring the largest possible variety of ecosystems for traces of the presence of molecules with biocidal properties. For example, a sample taken from a field in Missouri contained a bacterium later named *Streptomyces aureofaciens*, which, when put to work in the Lederle laboratory's fermentation tanks, produced astronomical quantities of the antibiotic Aureomycin. And Pfizer's Terramycin was apparently named after the site in Terre Haute, Indiana, where it was found in a lump of earth.[18] Just like sugar cane, microorganisms have been removed from their environments and put to work in systems with extremely simplified ecologies, fed with standardized culture media, and developed until they attain optimal productivity. The fruits of their labor have then been "delocalized," used in numerous ecosystems in which the bacteria that generated them did not necessarily belong. Tons of Aureomycin, streptomycin, Terramycin, penicillin, and other antibiotics have since been poured onto crops and livestock and into bodies, enabling the increase in the number and size of an ever-growing number of plantations by eliminating all beings deemed undesirable.

Plantations within plantations within plantations. This brings us to a terrifying story within a story: Simplifying ecosystems by eliminating interactions and eradicating certain living beings has led to the weakening of cultivated, raised, and produced species. Antibiotics have been used to counter the emergence of undesirable and pathogenic bacteria that could decimate entire crops and herds. However, this miraculous remedy has in turn generated new plagues and new feral species because although antibiotics are naturally excreted by bacteria, fungi, and other microorganisms, they have been produced on an entirely unprecedented scale since the 1950s, a transformation that has massively and permanently transformed the bacteria involved (see box).[19]

Plasmids—the circular DNA molecules on which resistance genes accumulate—play a fundamental role in the dissemination of antibiotic-resistance genes, leading the molecular evolutionist Michael Gillings to regard them as xenogenetic pollutants, "analogous to xenobiotic compounds [substances with toxic properties, even at low concentrations], but with the essential distinction that they replicate rather than degrade when released to pollute natural environments."[23]

An Increase in Evolutionary Forces in the Second Half of the Twentieth Century

- Increased natural and artificial mutation rates owing to the circulation of other mutation factors such as chemical mutagens, stress, and other antibiotics acting at the DNA level
- Increased selection pressures owing to the use of high concentrations of antibiotics and the speed of response to selection[20]
- Increased population size under selection pressure owing to the massive use of antibiotics in various ecosystems (very often with the same antibiotics used in both humans and animals, including the twenty-eight billion farm animals on the planet)[21]
- Increased migration rates of mutant strains owing to increased trade, enabling, even before the expression of their phenotypes, the dissemination of mutants that might otherwise have appeared and disappeared locally (by random effect or because of an excessively high selective cost related to the change in phenotype)[22]
- Increased intra- and interspecific genetic exchanges through the recombination or integration of various mobile genetic elements such as plasmids, which enable the concentration and accumulation of antibiotic-resistance genes

---

The bacterium *Acinetobacter baumannii* (better known as "Iraqibacter" because it has been responsible for many infections in American soldiers returning from Iraq, which partly explains the interest it has attracted) is an emblematic example. A metagenomic study published in 2006 revealed the presence of a resistance island of 86,000 base pairs in certain strains of this bacterial species, one of the largest described to date, encoding forty-five of the fifty-two genes for resistance to multiple antibiotics carried by the bacteria. One after the other, forty-five genes that enabled *Acinetobacter baumannii* to resist almost all currently available antibiotics were identified on the same DNA fragment. The analyses also showed that most of these genes came from bacteria belonging to the genera *Pseudomonas*, *Salmonella*, and *Escherichia*, thus testifying to numerous interspecies exchanges (i.e., between quite different bacterial species).[24] Since this study was conducted, the existence of strains resistant to all antibiotics, including colistin (which was usually the only treatment option despite its strong toxicity to the renal system but which has since been used in intensive pig breeding in Asia), has been proven.

Resistance is acquired sequentially with the introduction of new antibiotics to the market and then to farms, but, as shown by the case of Iraqibacter, it is also linked to other, sometimes unsuspected, factors. In a paper presenting an ongoing interdisciplinary research project, the anthropologist Omar Dewachi demonstrates how outbreaks of Iraqibacter cannot be dissociated from its place of origin—Iraq—and from a wartime context of collapsing health systems, the restriction of antibiotic use to a handful of molecules used as "quick fixes," and wounds highly conducive to bacterial infections, in addition to bombardments leading to the vaporization of heavy metals in the atmosphere, which are known to promote the acquisition of antibiotic resistance in bacteria.[25] Iraqibacter, as Dewachi informs us, cannot be dissociated from the ecologies of war and intervention.[26]

Hannah Landecker describes the integration of human histories into the biology of bacteria as the "biology of history" since bacteria have integrated human history into their very biology since the introduction of penicillin; more precisely, they have integrated human histories into their biologies as multiple manifestations of pluribiosis, situated encounters, and transformations. Consequently, "the bacteria of today are not the bacteria of yesterday, whether that change is registered culturally, genetically, physiologically, ecologically or medically. Bacteria today have different plasmids, traits, interrelations, capacities, distributions, and temporalities than those of bacteria before modern antibiotics. It is not even clear that 'bacteria' remains the only or the most salient category with which to think about antibiotic resistance. This biological matter, chewing away its own ontology, is historically and culturally—and materially—specific to late industrialism, produced in and by previous modes of knowledge."[27]

Antibiotics and plasmids are substances that leak, spread, and propagate, the miracle solution to myriad conditions that are becoming problematic, exposing the infrastructures from which they escape. This makes the "great acceleration" seem closer to a "great transformation": a profound mutation not only of certain relationships between humans and other human and nonhuman living beings but also of living beings themselves.[28] In this sense, a "deceleration," or drastic reduction in the quantity of antibiotics used—especially in agriculture and livestock—although absolutely necessary, would be far from straightforward, especially because it could not occur without completely transforming the dominant modes of production, the forms of which also vary according to context. The recognition of pluribiosis obliges us to accept such a requirement.

However, such a deceleration would be far from sufficient because presenting antibiotic resistance as a scourge that can still be contained is a vision of life that ignores what Michelle Murphy, in reference to the intergenerational effects of certain chemical molecules such as bisphenols, calls the "latency of infrastructures," that is, the time elapsed between a stimulus and the induced response. Bisphenols ingested by people decades ago are causing serious pathologies in first-, second-, and even third-generation offspring. Murphy tells us that because of this latency, *the future is already altered.*[29] In the case of antibiotics, both the persistence of resistance and its propagation—conveyed by the "xenogenetic pollutants" that are plasmids—are accomplished without the need for the presence of these antibiotics in the environment. This outcome has been known and exploited since the 1960s, when it was published in Anderson's reports.

This paints a very bleak picture because these pernicious ecologies, however simplified they may be, form complex ecosystems, tortuous entanglements, and unstable multispecies assemblies that must be individually analyzed and described. Alexis Zimmer provided such a description with regard to another type of pollution: the toxic mists that engulfed France's Meuse Valley in 1930. Unraveling and redefining the links among meteorological situations, industrial infrastructures, and the population, Zimmer presents the conditions—natural, technological, social, and economic—for the production of such fogs, but he also shows that these conditions have been maintained despite the established knowledge of their consequences. Zimmer's conclusion is inarguable: "In this regard, 'we' have not learned enough. In our interpretations of these phenomena, we have not learned to link the elements involved in pollution to our lifestyles, to the choices of societies, to the way in which these choices are collectively or not discussed and made, to the very unique types of relationships that our bodies weave with the air, water, land, the multitude of lives that are involved, and to the multiple historical trajectories that shape and inherit them."[30]

By putting "we" in quotation marks, Zimmer is making a distinction between a generic "we"—an abstract collective mobilized by the established and dominant knowledge—and the concrete collective, which could or should be established in a political reappropriation of the knowledge produced following the appearance of the fogs. However, this "we" can and must also be interpreted in relation to a territory, especially given that, as Zimmer tells us, the industries at the origin of this pollution have not "disappeared"; they have simply been moved from Belgium to China and subsequently expanded. The similar trajectory of many other types of production, including antibiotic-intensive factory farming, is a reminder of the deeply rooted and historical links between pollution and colonialism.[31]

While I do not purport to provide solutions to the problems triggered by generalized antimicrobial resistance—this "intrusion," as Isabelle Stengers puts it—I do assert that it is a product of modes of relationships and production that are ignorant of pluribiosis. Responses to such an intrusion cannot and must not be allowed to replicate the conditions in which the disaster occurred. They must address both its consequences *and* its causes.

The implementation of solutions, whether by developing new antimicrobial agents or therapeutic alternatives, therefore requires reflection on the materialities specific to both therapeutic entities and capitalist infrastructures, as well as the requirements for the scalability of the production and exploitation of living organisms that underlie them, including in terms of the complexity of the relations between countries of the Global North and Global South on which they are based. There will be no point in producing new antibiotics without thoroughly rethinking our conceptions of them, their action, and their interactions. Similarly, bacteriophage viruses can be used to treat bacterial infections that have become antibiotic resistant—but only subject to conditions that will also prevent them from contributing to the causes of widespread bacterial resistance.

9

TOWARD A PLURIBIOTIC MODEL?

What response can bacteriophage viruses provide to the intrusion of bacterial antibiotic resistance? There is no simple answer, no miracle recipe for "controlling" or "mastering" living beings, no formula for "taming" certain entities—living or otherwise—so that they respond, without recalcitrance, to the projects that seek to harness them. Can viruses be used as therapeutic entities? That was the initial question. This issue is now encapsulated in the complexity of the various entanglements in which phages, bacteria, and humans are caught up, in the intricacy of the various scales that must be considered when striving to understand the actions and reactions of these living entities, from the microgeohistories constituted by isolated bacterial infections located in the bodies of patients to the propagation without borders or barriers—whether spatial or of species—of multiresistant plasmids.

Formulating a response will therefore be particularly challenging, requiring consideration of not only the agency of the various entities but also human infrastructures and their materiality. I do not claim to provide such a response, which can only be collective, complex, and multifaceted. However, it is possible to outline general frameworks for a response, informed by all the

176 • TOWARD A PLURIBIOTIC MODEL?

people we have encountered so far: those working in laboratories, hospitals, and regulatory agencies; those involved in creating collections, treating patients, and developing protocols for clinical trials; and those shaped by how these people devise and use relationships with microbes, as well as the knowledge they produce, mobilize, and disseminate. Throughout the preceding chapters, I have described the production of knowledge by chronically infected patients like André (chapter 1), in microbiology and microbial ecology laboratories (chapters 3 and 4), and on infectious disease wards (chapters 5 and 6), showing that knowledge production occurs concomitantly with the configuration and conceptualization of new relationships with microbes, which also simultaneously reveals profound and much older relationships. The notion of pluribiosis reflects the spectra of relationships involved and the many ways that encounters transform living beings, as well as the situated and engaged character of the knowledge that objectifies those beings. By developing their understanding of the new microbial relationships they have learned to see, the experts I encountered are building an "art of attentiveness" that translates into *moral and political* propositions concerning how bacteriophage viruses should be used, thus establishing a political ecology of microbes.[1]

Our response to antibiotic resistance must include these moral and political propositions, right down to the way in which the rails needed to give phage therapy an existence in society (here in France) are built and laid. In other words, right down to the development model(s) that are or will be put in place. As the philosopher Émilie Hache points out, "The accumulation of scientific data no more eliminates moral concern than the latter can dispense with the development of closer ties with the beings with which it is concerned. Accepting that morality should intervene in science, in this sense, amounts to revisiting the

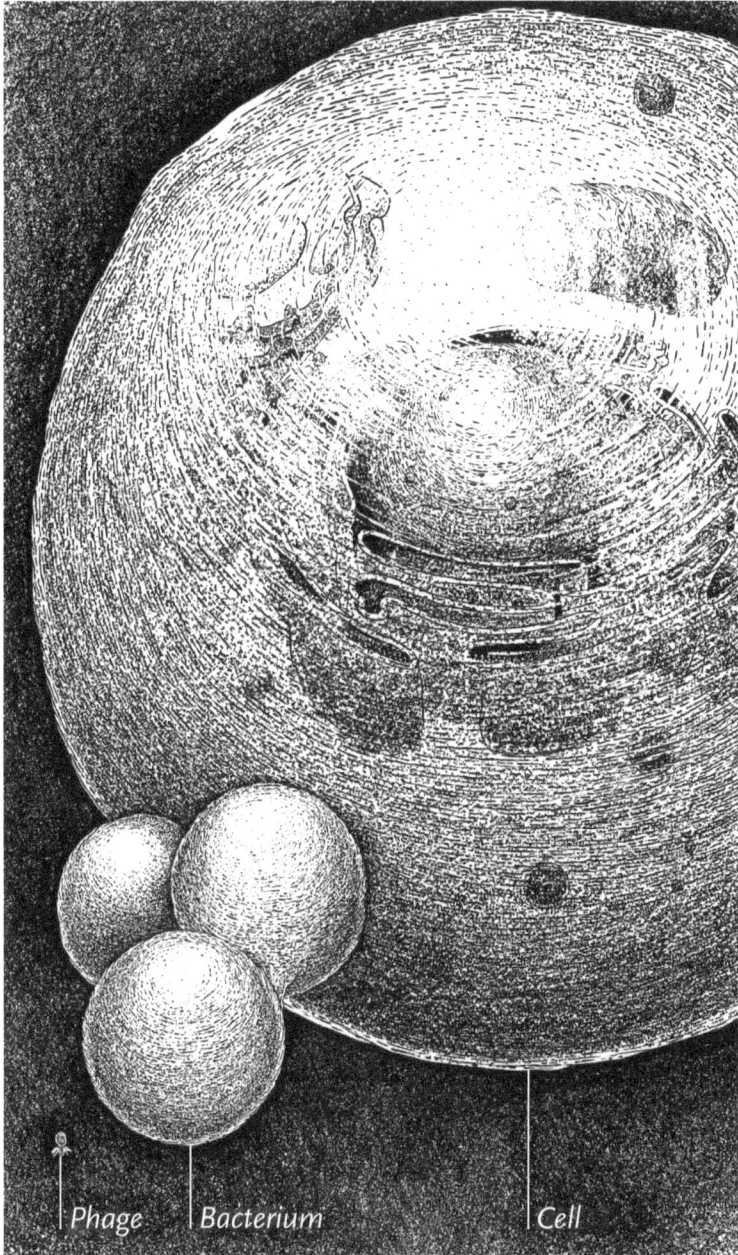

FIGURE 9.1 A phage, a bacterium, and a eukaryotic cell

interference of different causes, with 'intervene (in)' not meaning 'take the place of' but rather *hold together* inseparable dimensions of being."[2]

Because these propositions are conveyed by a variety of initiatives and are reflected in stances taken repeatedly over the years, they expose the flaws and ambiguities described in the previous chapters. However, to reestablish the power of these propositions and their expression, it was first necessary to unpack all the relationships they are supposed to hold together and bring to life. In this final chapter, which will set out the development models in which phages exist or could exist, I will attempt to explain how "the consideration of these moral demands, by requiring us to ensure the decent collective treatment of as many beings as possible, leads us to rethink the *political* make-up of our societies."[3]

A REGULATORY CONUNDRUM

Members of the Queen Astrid Military Hospital team in Brussels were among the first people I met for this research. Drawing on their wide-ranging expertise, this multidisciplinary team (composed of Maya Merabishvili, Jean-Paul Pirnay, Mario Vaneechoutte, Gilbert Verbeken, Daniel De Vos and Martin Zizi) initiated the 2011 paper titled "The Phage Therapy Paradigm," which proposed the development of sur-mesure treatments. In addition to their research on phage isolation and purification, their participation in clinical trials, and their involvement in treating patients, this team has for many years been a regular contributor to conferences and workshops and to publications in the specialized and general press, promoting the development of therapeutic phages and a specific approach to their use in the context of infections: a sur-mesure approach adapted to the unique needs of each patient and accessible to all.[4] However, as they tell us, this work entails more than simply producing data. It also requires the creation of a space where such a conception of patient care can actually exist, that is, a *regulatory framework specific to phages.* Indeed, as we saw in chapter 2, despite all the scientific, technical, and medical work carried out by the Queen Astrid team, by the Georgian scientists in Tbilisi, by teams in Poland and Russia, and by teams in American, British, and French laboratories and hospitals, phage therapy has no validity in Europe—because bacteriophage viruses have no regulatory existence.[5]

The Queen Astrid team therefore set about this task in the first decade of the twenty-first century in what turned out to be an unsuccessful attempt, which they recounted in the evocatively titled 2007 article "European Regulatory Conundrum of Phage Therapy."[6] The reasons given at the time for being unable to have phages considered a potential therapy show the progress made in the past fifteen years, as well as the profound changes that have rendered that judgment totally obsolete. Back then, viruses were considered dangerous (the same reason given for rejecting Alain Dublanchet's propositions, as discussed in chapter 2), and antibiotics were still largely effective.

In 2011, however, in view of the growing interest in phage therapy and the inescapable problems caused by antimicrobial resistance, the European Medicines Agency (EMA) and the US Food and Drug Administration (FDA) deemed that phages were medicinal products and would therefore be required to meet current norms and standards.[7] This qualification would be confirmed four years later, during a workshop held at the EMA on June 8, 2015. However, most of the international specialists present at this event (including doctors, microbiologists, pharmacists, and specialists in phage–bacteria interactions) protested a few days later in a letter in which they repeated their call for the creation of a specific regulatory framework that would, in their view, make up for the lack of public and private investment and develop the sur-mesure approach.[8]

The refusal to equate phages with medicinal products, expressed by most people working on phage therapy, may come as a surprise. After all, phages now have regulatory status and can be developed to treat people with bacterial infections. The actors involved, including scientists, sometimes justify their disapproval by referring to the norms and standards associated with drug production, which are indeed highly restrictive, and the need for qualified facilities that would be recognized by the National Agency for the Safety of Medicines and Health Products (ANSM) or the EMA. The authors of the letter (published in the journal *FEMS Microbiology Letters* in 2016) insist that the "consensus" imposed by the EMA prevents researchers from taking the evolving capacities of phages and bacteria into account and therefore from regularly adapting treatments in response to that evolution. It should be kept in mind that because of the extreme specificity of the relationships between phages and bacteria, compliance with existing regulations and with good manufacturing practices (GMP) would require the adaptation of production and purification processes for each new phage, as well as an assessment of its efficacy and nontoxicity, in one or more clinical trials.[9] Yet all those involved consider the production and purification of phages and the acquisition of data, whether through animal models, compassionate care models, or clinical trials, to be essential. Everyone is working tirelessly to build and characterize phage collections and their specifications, which, as we have seen, are specific to each virus. All of this data is necessary for safe, high-quality production. Similarly, while clinical trials, because of their standardizing and universalizing nature, are often held up as an example of the inadequacy of the regulatory framework, we have seen that it is possible to design such trials differently, to ensure they account for the sur-mesure nature of phage therapy. In other words, although opponents of the categorization of phages as drugs present technical and scientific explanations as evidence of the inadequacy of the regulatory status vis-à-vis the phage therapy project, it is not so much these imperatives that are problematic as *how* they will be implemented, for *what purpose*, and therefore *by whom*.

Pharmaceutical companies are entirely responsible for the regulatory category of drugs. The various standards and norms to be met relate to one goal: obtaining the precious marketing authorization. And GMP require dedicated plants and approved pharmaceutical production facility status.

BIG PHARMA

If Big Pharma has been mentioned only once so far in this book, it is not for the sake of maintaining suspense. It is simply because pharmaceutical companies are conspicuous by their absence from phage therapy developments—an absence with far-reaching consequences. In early 2018, Gilbert Verbeken told me that phages still lacked regulatory existence at the end of the first decade of the twenty-first century because "pharmaceutical companies saw no interest in them. It's not what the industry is used to doing. The industry is more general; it's all about *market placement*, uniform products, things like that. Phages evolve, bacteria evolve—it's all very dynamic, and if you really want to use phages to treat patients, it will require a lot of hard work."[10] The European and national regulatory agencies do not produce specific regulations for the therapeutic use of phages *because pharmaceutical companies do not see any point in it.* "The authorities aren't there to produce regulations; they're there to enforce them," Verbeken told me. Therefore, it is up to those who want to introduce phage therapy to make it happen, to demonstrate its importance and how it should be practiced. This is the context in which we should understand the numerous papers written and appeals launched by the Queen Astrid team, soon to be joined and supported by an ever-growing community of researchers.[11]

While pharmaceutical companies remained on the sidelines, start-ups began working on phages and started lobbying to bring phage therapy into the regulatory arena. The turn of the 2010s saw the start of a power struggle for the emergence of one model over another, which led to the current categorization, which suits only private actors and is clearly in keeping with the spirit of the times.

Today, it is almost impossible to consider the question of drug development outside Big Pharma networks and infrastructures, and with good reason: Decades of transformations and the use of scientific metrics to link health and disease issues to market concerns have led to "global health [being seen as] as a market- and profit-driven opportunity."[12] We now live in what the anthropologist Kaushik Sunder Rajan calls a "pharmocracy": a "global hegemony of the multinational pharmaceutical industry," historically made possible by two waves of harmonization. First and foremost, the extension of regulations on clinical trials— to comply with the guidelines established by the International Council for Harmonisation of Technical Requirements for Pharmaceuticals for Human Use—has created a global market with ever-increasing outlets for new molecules to be brought to market in the Global North. However, it has also encouraged the offshoring of clinical trials and, with that, the production and invisibilization of cheap labor based on the exploitation of

populations considered subservient.[13] A form of mass production and consumption has been made possible by the uniformi-
zation and standardization of ailments, diseases, measuring
instruments, physiologies, bodies, and psyches.[14]

But these are not the only reconfigurations. The harmonization of drug regulations was necessary to extend the use of clinical trials and increase the mass of new prescriptions. However, a
harmonization of intellectual property rights was also required
to secure the investments made by pharmaceutical companies.
In a profit-driven approach, what is a molecule worth if it can be
copied, produced, and sold at prices lower than those charged by
the proprietary firm in countries that do not recognize the patent laws in force in the United States, for example?[15] The hegemonic regime of the pharmaceutical industry varies according
to national context but depends largely on the profound asym-
metries between the Global North and the Global South, as well
as on the maintenance, in constantly renewed and contingent
forms, of certain systems of domination.

In the second half of the twentieth century, the pharmaceutical industry's never-ending drive for profit led to the abandonment of entire health care sectors. The term "neglected tropical
diseases" refers to pathologies affecting populations that lack the
resources required to constitute attractive markets. However,
this prioritization of health issues is not limited to the Global
South. Indeed, for several decades, the pharmaceutical industry
has been focusing primarily on treatments for chronic diseases,
which are becoming increasingly common because of longer
life expectancies and lifestyle changes.[16] However, this increase
has been to the detriment of anti-infective treatments, particularly antibiotics, which were developed at great cost and are
now coming under scrutiny because their efficacy may rapidly
diminish.[17] Since the early years of the twenty-first century, the

ever-increasing financialization of the pharmaceutical industry has led companies to focus on short-term merger-and-acquisition objectives while neglecting research and development (R&D). This has led to a shortage of new molecules for treating even common ailments.

This chilling observation helps explain pharmaceutical companies' lack of interest in phage therapy. As Alain-Michel Ceretti pointed out when recounting his attempt to interest manufacturers in the initiatives of the start-up Pherecydes Pharma (as discussed in chapter 2), "They're not going to follow this path [of phage therapy] because nothing is patentable. And that's when I started to understand the problem. In fact, previously, the key issue was not patentability; it was usage, importation, or manufacture. In the meantime, it has become a medicinal product. Manufacturing it has become complicated, and importing it from a country where there are no clinical studies is impossible."

As we have seen, the work required to standardize knowledge and know-how, mechanisms, methods, and standards, combined with the epistemological and ontological importance placed on the administration of evidence, makes clinical trials not only fascinating devices but also particularly long and costly undertakings. From the pharmaceutical industry's standpoint, and therefore from the perspective of profitability, the difficulties on the intellectual property front are multiplied by the unique characteristics of bacteriophage viruses. However, I am less categorical than Alain-Michel Ceretti. If the question of patentability crops up so regularly in the opinion papers of major scientific journals, and in blogs, forums, and workshops devoted to the use of phages, it is because the issue has not yet been settled.[18] And it is easy to understand why since the stakes are so high: Finding a way to patent phages is the key to securing a proportion of the substantial investments required to build collections and set up clinical trials—but the situation is far from simple.[19] Although it is hard to separate the legal and moral question of patenting living beings from the material and technical aspects of this work, I can shed light on the latter.[20] A bacteriophage virus placed in a collection is a snapshot of a microgeohistory: It has been isolated from water in a specific location, often from a bacterium taken from a sick person in a hospital or laboratory by a technician who has altered numerous experimental conditions to ensure the best possible "communication" between the virus and the bacterium. Such a phage, placed in contact with another bacterial strain or with the same strain but in a different medium, is likely to evolve relatively quickly. The question that then arises in the context of patenting is quite simple: How do we define, and how can we define, the "identity" of a bacteriophage virus?

Clinical trials of phages, whether sur-mesure or prêt-à-porter, will be long and costly. In addition, it will be extremely difficult to obtain any return on investment since patenting a phage is always an arduous task, given the pluribiotic capacities of living organisms. These two aspects—the cost of clinical trials and the uncertainty surrounding the possibility of patenting phages and thus of securing major investments—explains Big Pharma's apparent lack of interest in this issue. At least *for the time being*.

Drug status is therefore problematic because it makes phage therapy an initiative that can be fully developed only by the pharmaceutical industry, whereas it is public laboratories, hospitals, and start-ups that have been primarily responsible for its existence.

While the majority of experts I have met in recent years insist on the need to change the regulatory status of phages, it is not necessarily because they want to attract more private investment; far from it.[21] In this case, the lack of pharmaceutical company involvement—which is considered an obstacle in the *dominant drug development model*—represents an opportunity to put forward propositions that are diametrically opposed to Big Pharma's priorities.

PROMOTING THE EMERGENCE
OF OTHER PROPOSITIONS

Supporting sur-mesure development that is adapted to micro-geohistories; ensuring that phages are not treated as yet another antibiotic while facing the attendant risks of destroying commensal or symbiotic microbial populations and triggering the emergence of bacterial resistance to these viruses; avoiding the indiscriminate use of phages on livestock and crops, as occurred with antibiotics; creating new practices and new knowledge; patiently developing prudent uses for phages while acknowledging pluribiosis and its diverse manifestations; developing a medicine that is mindful of patients and their particular needs; creating opportunities rather than reducing them; protecting bacteriophage viruses from market logics.

Is it better to amputate patients' limbs and fit them with prostheses or to allow them to live with the bacteria that are slowly eating away at their bones? Is it better to administer an antibiotic, the renal toxicity of which will eventually deprive sick people of functioning kidneys, or to administer viruses? If phages are produced by profit-seeking manufacturers and then patented and sold at a premium, the answer to these questions will be unequivocal and profoundly unjust: The choice will depend solely on the economic capital of the sick people or of their country's health care system.

Is it better to administer a cocktail or a treatment customized to a patient's particular infection? If cocktails are formulated to be used as widely as possible, if they are designed to mimic the broad-spectrum effects of antibiotics, and if they follow the trend of the latter, then the response may well lead to further disasters. Beyond the very real risks associated with the development of bacterial resistance to the bacteriophage viruses contained in these assemblies (as discussed in chapter 5), no one is yet able to predict the effects of a massive release of bacteriophage viruses into a given environment. And if bacteriophage cocktails are effective against pathogens carried by humans, why should they not also be effective against those we share with animals? The history of antibiotics, their uses, and the emergence and interspecific transmission of bacterial resistance is a cautionary tale. Everything about phage therapy is highly uncertain. Perhaps, in ten or twenty years' time, cocktails will be created and used in specific cases to treat specific ailments, which could reduce both the workload of health care personnel and treatment delays. Perhaps they will be useful. Nothing is ruled out; nothing is set in stone. The questions I have posed here are not simple, but phages provide the opportunity to raise them. That is what the people I have encountered in my research want: to keep the possibilities alive.

The Magistral Phage

The team at Queen Astrid Military Hospital has proposed an interesting and elegant side step by taking advantage of the great complexity of drug regulation and lobbying to put phage therapy on the Belgian government's agenda. While there are no specific regulations at the European level, there is consensus on the categorization of phages as medicinal products. It has thus been proposed that phages be used as ingredients in magistral preparations. The European regulations define a *magistral prepa-ration* as "any medicinal product prepared in a pharmacy in accordance with a medical prescription for an individual patient."[22] A magistral preparation is made using two types of products: (1) *authorized ingredients*, that is, those listed in the national or European pharmacopoeia (new ingredients may be added to a pharmacopoeia after consultation with the National Commission and, in Belgium, after approval by the Ministry of Public Health), and (2) *unauthorized ingredients*, that is, those not listed in the pharmacopoeia but which may be used if their production complies with a specific monograph (a type of specification for high-quality biological production meeting requirements set by regulatory bodies) and if their biological quality is certified by an independent laboratory, which may be public or private.[23]

To account for the enormous diversity of phages, which is all the more necessary when supporting a sur-mesure strategy, the Belgian government has decided to classify these viruses as unauthorized ingredients. This classification enables phage collections to be used according to a fairly simple procedure: When an infected person is examined, a phagogram is performed on the pathogenic bacterium to determine and quantify the activity of the phages in the collection on that strain. Only active phages are produced, according to the monograph, and sent to the certification laboratory for validation. They are then administered under the responsibility of the prescribing physician and pharmacist.[24] The reclassification of phages as unauthorized ingredients is accompanied by a government commitment to reimburse the cost of phage therapy treatments for a two-year period, after which a reassessment of progress made will be conducted.

In response to the influx of requests for treatment (partly fueled by several reports on Belgian and Dutch television channels), the Queen Astrid team asked the government for greater financial investment so that it could hire staff to build up their collections and increase their production and quality control capacities.[25] In response, rather than increasing public investment, the government proposed to develop a partnership with private companies to outsource some of the work. There is no telling which party will benefit from the development of this public-private partnership in the years to come.[26]

The change in regulations for phages is of the utmost importance as it opens up new opportunities. The magistral preparation model facilitates a rapid response to the specificities of each infection, accounts for the evolutive capacities of microbes, and facilitates the type of pluribiotic approach sought by the people I have interviewed. However, there is no guarantee that it will last, as I came to understand when I became involved with an initiative at the Hospices Civils de Lyon.

The "Phage in Lyon" Project

In spring 2019 at the Croix-Rousse Hospital in Lyon, I met with Tristan Ferry, the deputy head of the infectiology department, and with Frédéric Laurent and Gilles Leboucher, respectively the heads of the bacteriology laboratory and pharmacy department. We had previously crossed paths at various workshops, and their activity in the phage therapy field was beginning to gain recognition owing to the number of people they had treated and their effective communication strategy. On March 21, 2019, the three department heads and I were participating as experts at the second Temporary Specialized Scientific Committee (CSST) on phage therapy, organized by the

ANSM (as discussed in chapter 1). The purpose of this committee meeting was to take stock of the situation and assess the progress made since the first CSST, which was held in 2016 and had noted the need to actively support the development of phage research.[27] The aim of the second CSST was "to illustrate the types of phages, patient treatment strategies, phage selection methods and routes of administration" that should be used in phage therapy.[28] From an ethnographic standpoint, this committee provided an excellent opportunity to observe the interlinking of various approaches—whether medical, microbiological, technical, economic, political, or financial—in the creation of a new therapy. For several hours, a series of experts reported on their activities and the successes, failures, and difficulties they had encountered, such as the need to test the equipment designed to contain phages (since phages adhere to certain materials, thus compromising their administration to patients) and methodological issues relating to the determination of phage activity, which plays a crucial role in the choice of viruses to be used.

In its conclusions, the CSST stressed the need to obtain proof of efficacy, and therefore to develop clinical trials as quickly as possible, since case collections do not meet current regulatory requirements: "This case-by-case use of phages for which research and development activities have been conducted enables the formulation of responses to specific requests, but cannot replace the need for efficacy and risk data from clinical trials in the field of phage therapy. The very low number of clinical trials in this field to date perpetuates the uncertainties and questions concerning the degree of efficacy and the administration procedures according to the therapeutic objectives."[29]

The main problem, however, was—and remains—the lack of available phage preparations needed to develop clinical trials, especially those produced according to GMP. At the second CSST, representatives from Pherecydes Pharma stated their position in favor of producing not cocktails but individual phages, and they presented their remarkable advances in phage isolation, production, and qualification.[30] Four anti–*Pseudomonas aeruginosa* phages and three anti–*Staphylococcus aureus* phages were in the running for future temporary use authorizations as soon as their GMP production was approved. These phages could then be used in partnership with public hospitals to launch clinical trials in various pathologies (e.g., osteoarticular, diabetic foot, respiratory, and complicated urinary tract infections). Pherecydes Pharma was expected to submit a marketing authorization application for an anti-staphylococcal phage by 2023.[31]

The CSST also praised the work being carried out in Belgium on increasing access to phages and stressed "the importance of also having a phage library and local academic production. In this respect, the ANSM has always emphasized the importance of having academic operators alongside industrial operators, in order to diversify and expand the national phage production offering in France."[32] The ANSM added, "To date, there is [was] no equivalent in France of academic production derived from a phage library as carried out according to quality standards at the QAMH [Queen Astrid Military Hospital] in Belgium."[33]

In view of these factors, the ANSM asked CSST members for their opinion on the merits of creating a "national platform for guidance and validation of the use of phages," which may also be used as a platform for "academic production" (box).

Objectives of the National Phage Therapy Management Platform as Presented in the 2019 CSST Minutes

- Confirm, in a collegial manner, that the patient's clinical situation and treatment history justify the use of phages.
- Investigate the availability of phages according to the species responsible for the infection.
- Work toward the implementation of academic production in France based on a phage library along the lines of the one established according to quality standards at HMRA, in order to broaden the healthcare provision and promote a dynamic approach to bacteriophage research at national level.
- Obtain a collegial analysis centralized at the academic level, with members possessing microbiological expertise for the interpretation of phage activity data, given the lack of references and limited hindsight for the analysis of these data.
- Provide guidance on the most appropriate use of phages in the clinical context, based on the available data.
- Establish links between the use of phages on a compassionate-access basis versus inclusion in trials (in line with always favoring early access via clinical trials).
- Contribute to the analysis of benefit and risk data.
- Provide healthcare professionals and patients with up-to-date information on the use of bacteriophages in hospitals, based on the current state of knowledge.[34]

The ANSM representatives added that this platform should be set up at the ministerial level with the bodies involved in the organization of the health care system. In other words, responsibility for the effectiveness and operability of such a platform was referred to France's Directorate General of Health and Directorate General of Health Care Provision. In its half-day meeting, the CSST provided an insight into the enormity of the task that lies ahead. Everything, down to the injection equipment used to administer phages, must be developed, tested, and proven.

The CSST also provided an opportunity for lively discussions.[35] Tristan Ferry, Frédéric Laurent, and Gilles Leboucher were all present to report on their activities and their complementary aspects: Ferry presented several cases of complex osteoarticular infections that he and his colleagues had treated in collaboration with Sébastien Lustig, the head of the orthopedics department at Croix-Rousse Hospital, as well as a case of endocarditis (for a total of seven people treated with phages in two years). Laurent described the pharmacological studies his team were conducting on these cases (pharmacokinetic follow-up), and the microbiology research being conducted on the potential impact of bacteriophages on bacterial biofilm. Leboucher presented his expertise in phage administration. These presentations seemed to suggest the emergence of a major research center in Lyon.

One month later, I spent more than five hours chatting with Tristan Ferry, during which time he reviewed the last few years: how phages had gradually emerged as one of several answers to the problems encountered in the management of antibiotic-resistant infections and how everyone at Croix-Rousse Hospital, enthused by the opportunities provided by bacteriophage viruses, by the scientific, technical, and medical challenges they posed, and by the need to make progress, had committed themselves fully (and legally) to this adventure. They had reached the same conclusion as at the CSST: We need phages. Isolating them is not the hardest part. Above all, it is a question of guaranteeing high-quality production to enable phages to be used in humans. The day that I was speaking with Ferry, he gave me some information that was not yet official: The Hospices Civils de Lyon had established a refurbished production platform capable of meeting pharmaceutical production standards. Consequently, the combined competencies in microbiology, infectiology, pharmacy, and surgery available at Croix-Rousse Hospital were now accompanied by the opportunity, albeit nascent, to produce phages that could meet the highest standards of pharmaceutical production.

Three years and dozens of treatments later, Lyon now has an environment in which phage therapy can be designed and delivered.[36] All steps of the process—isolating phages in wastewater; developing high-quality purification and production procedures; creating in vitro, ex vivo, and in vivo models to study the many interactions among phages, bacteria, cells, organs, and environments; modeling pharmacokinetic and pharmacodynamic data; designing and developing clinical trials; and administering phages—are now in various stages of development.

DEVELOPMENT MODELS
AS POLITICAL ECONOMIES

Lyon has now established the equivalent of the care-and-production platform called for at the 2019 CSST. At that meeting, however, an ambiguity was pointed out that persists to this day, the same one that has preoccupied us since the beginning of this book: the regulations to which phages must conform. Although the ANSM and all those present at the 2019 CSST have called for the public production of phages (notably to compensate for the lack of interest shown by private companies), it is unclear what status such production should have. Without going into technical details (which are of little interest at this stage given that even the specialists disagree on the interpretation of regulatory texts), it is important to stress one significant point on which everyone agrees: the preeminence of the private sector over the public sector in drug development. As long as there are no commercial preparations of bacteriophage viruses, health care professionals remain free to make use of the public collections used for magistral preparations. They are even encouraged to do so. However, since hospital pharmacies, like public research laboratories, do not have approved pharmaceutical production facility status, these collections cannot be qualified as GMP compliant, even if their production meets equivalent standards.[37] This seems to imply that the phages in public collections are to be used only in a compassionate use context, with the prescribing physician and the administering pharmacist incurring full responsibility for the patient's treatment—unless France adopts Belgium's regulatory framework.[38]

But even if France does so, the existence of "academic" phages will remain precarious: The cohabitation of two production models—public and private—cannot last long given the current state of the law. As soon as commercial preparations become available, health care professionals will be legally obliged to use them.[39] The expression "diversify and expand the national phage production" takes on a very different meaning here from what it might have had on first reading. In this story, public research plays the same role that it has repeatedly assumed in the health care field for decades: bearing the R&D costs and doing most of the work before the results are, at best, shared within public–private partnerships or, at worst, confiscated outright by industrial operators, increasing the risk of denying the propositions highlighted in this book an opportunity to exist.[40]

However, the dominant proprietary model of the pharmaceutical industry need not be inevitable. Although we are clearly confronted with a hegemonic system, alternatives do exist, based on the consideration of existing but often invisible relationships.

An excellent example of this is the struggle waged by people with HIV, and by certain governments, to gain access to anti-retroviral drugs at the height of the AIDS epidemic. In that case, by appropriating knowledge that had partly been denied to them, people with HIV were able to challenge regulations that hindered access to therapies still being evaluated in clinical trials. This enabled them to pressure the FDA into issuing marketing authorizations quickly and thus enabled thousands of patients to receive immediate treatment.[41]

Maurice Cassier's studies provide a particularly valuable analysis of the problems posed by financial capitalism and by this type of alternative construction, with a view to reducing the influence of exclusive markets. They show how development models, including the pricing of therapies, encapsulate economic policies: "They sum up the politics of appropriation of these molecules, with a view to obtaining a monopoly or a model based on the common good; they are related to economic value regimes intended to optimize the profitability of invested capital or to increase the accessibility of drugs for public payers and patients; and they are justified or contested by moral economies."[42]

Some initiatives have entailed the proposition of radically different models. Countries such as Brazil have challenged and overridden the systems governing the appropriation of antiretroviral molecules, which formerly prevented any organization other than the patent owner from producing those therapies. By pitting the health of populations, and therefore the common good, against intellectual property, it has enabled the development of new alliances and partnerships between public and private laboratories and ensured the production of these molecules on a larger scale at lower cost.[43] The Drugs for Neglected Diseases initiative (DNDi), an independent nonprofit research organization based in Geneva, aims to develop drugs to combat tropical diseases that are often neglected as a result of the same developments in pharmaceutical capitalism.[44] DNDi has developed a partnership with the pharmaceutical giant Sanofi, which has led to the invention, production, and distribution of a new fixed-dose combination of artesunate and amodiaquine (called ASAQ) to combat malaria. Thanks to a contract between Sanofi and DNDi—the former relinquishing the right to file a patent and the latter bearing a large proportion of the R&D costs, especially for clinical trials—this therapy was brought to market in 2007 by Sanofi at a price of $1 per treatment for adults and $0.50 for children.[45]

By considering phages to be drugs and, in the guidelines of various international organizations, as yet another antibiotic, the current regulations appear to position the development of this therapy within a proprietary model.[46] Pharmaceutical companies' lack of interest in the potential of phages—the raison d'être of this book—provides an opportunity. Our focus should be on collectively devising a development model that will enable the assertion of the positions and propositions described in the previous chapters in the various arenas in which the use of these viruses is conceptualized and constructed. A model that guarantees, over the long term, a reasoned use of these viruses in the closest accordance with their diverse microgeohistories. A model that accounts for the many potentialities of phages and the roles they may play in the various ecosystems in which they participate.

The Lyon platform and the initiative developed by the Queen Astrid Military Hospital team are concrete examples of such a model, originating from the aspirations of the people involved and the principles that guide them, as well as from the pooling of knowledge, know-how, and competencies that are constantly being developed, driven by individuals, teams, and institutions.[47] What is more, active collaboration among teams—as illustrated by the efforts made to find an active phage to treat a sick person described in chapter 3—provides a glimpse of the potential strength of an initiative based on a network of laboratories and public hospitals at the European level.

The aim, therefore, is not only to support these alliances but also to strengthen and protect them. This means fully embracing the work involved in producing and reproducing the facts but not invisibilizing those doing the work; considering phage collections not as reservoirs of potential wonder drugs, which they are not, but as snapshots of microgeohistories; highlighting and accounting for the time and attention required to treat patients' infections without making mistakes, including with regard to nonpathogenic bacteria; designing and developing clinical trials that will be all the more likely to demonstrate the safety and efficacy of phage treatments if the forms of division of labor on which they are based are accounted for, and all the more likely to meet their objectives if the health they may improve is not to the detriment of those working on it; establishing regulations and laws strictly governing the use of bacteriophage viruses and subjecting any extension or proposed new use to a multidisciplinary committee whose members must not represent interests other than those of the patients and ecosystems likely to be modified by these new practices; securing sustainable funding to enable the continued production of relevant knowledge independently of the pursuit of profitability and "valorization"; sharing the new knowledge that is generated—open sharing remains the best protection against privative appropriation; and offering phage therapy to all those who need it without discrimination based on class or race.

The initiatives and fledgling collectives working on phage therapy must therefore be nurtured, further developed, and linked to other groups doing work in the field. An infectious disease specialist at the Bordeaux University Hospital once told me that he was pleased to see me taking an interest in his practices and questioning the knowledge produced about the relationships between humans and microbes, as well as the mobilization and articulation of this knowledge. Unfortunately, his work, which was becoming increasingly burdensome because of the dismantling of the French health care system, did not give him the opportunity to engage in knowledge-sharing activities, even though they were essential. This observation was shared by all those I encountered, who were always enthusiastic to share their research, welcome me into their laboratories, and share their hopes and fears. Social sciences, more precisely a particular practice of ethnography—one that claims to be militant, is based on the strict requirements of feminist science and technology studies, and recognizes that ideas and knowledge are never innocent—can not only describe and analyze this knowledge but also create spaces to facilitate the convergence of scientific, regulatory, political, economic, and social perspectives and issues. In this way, it can unite relationships and propositions to create futures other than the mortifying prospects promised to the maximum number by the constantly renewed forms of capitalism as a mode of relationship.

It is a safe bet that, over time and as antibiotic resistance continues to grow, start-ups or pharmaceutical companies, or perhaps both, will eventually develop and market prêt-à-porter phage cocktails. They will offer a scalable product that is mass-produced, despite all the indications that it should not be, whose environmental effects cannot be predicted, and which will only partially correspond to the expectations and practices of people with a more ecological and dynamic conception of the relationships between humans and microbes. In so doing, they will continue to produce and perpetuate plantations.

The problems that demand the attention of humans, whether caused by global warming or chemical pollution, are multidimensional. They are the result of relationships among various types of human and nonhuman entities; complex infrastructures that are usually invisibilized; local and global regulations; economic systems; agricultural, health, and safety policies; and the exploitation of constantly renewed forms of labor—all of which are also ways of maintaining relationships. Each being exists through the relationships it maintains on various scales that are sometimes difficult to reconcile. Everything boils down to relationships. But not every relationship is a good one.

Antimicrobial resistance is a globalized phenomenon embodied in a situated, pluribiotic way. This means that no one-size-fits-all solution will ever be satisfactory. Indeed, any solution that purports to be universal should be viewed with the utmost caution; we should ask what relationships, what histories, and what scales have been disregarded to achieve such a consensus? Under certain conditions, bacteriophage viruses may provide an answer. With their multiscalar and multirelational histories, bacteriophage viruses, because of their very mode of existence, cause humans to hesitate. But each hesitation can be seen as an opportunity to provide a response that both addresses the consequences of antimicrobial resistance and accounts for its many causes. It is an exacting response that is needed, one that can never be taken for granted and that requires the establishment of power relations and the creation of and support for collectives that fight tirelessly to protect the interests of the maximum number of people within the damaged ecologies that humans inherit. A response that is compatible with Gaia and pluribiosis and therefore struggles against the stranglehold of capitalism.

ACKNOWLEDGMENTS

A book is a snapshot of ongoing research, a recording on paper of the ideas and diverse environments in which they continuously evolve and involve through encounters and interactions. These acknowledgments only partially do justice to those who have accompanied and nourished this work, some for a long time, some intensely in recent months.

This book would not exist without the people who populate it and all those who allowed me to discover and explore the rich and surprising world of bacteriophages. I sincerely thank Alain Dublanchet for his time and trust, given repeatedly, and for his careful reading. Rémy Froissart, to whom this work owes much, and Claire Le Hénaff-Le Marrec have accompanied me from the beginning of the investigation, as have other members of the Phages research group: Mireille Ansaldi, Pascale Boulanger, Laurent Debarbieux, Nicolas Dufour, Sylvain Gandon, Marie-Agnès Petit, Eduardo Rocha, and Clara Torres-Barcelo. The enthusiasm of the "phagists" has never wavered, through meetings and discussions with Leslie Blazere, Alexandre Bleibtreu, Alain-Michel Ceretti, Frédéric-Antoine Dauchy, Raphaëlle Delattre, Tristan Ferry, Patrick Forterre, Camille Kolenda, Floriane Laumay, Frédéric Laurent, Gilles Leboucher, Tiphaine

Legendre, Aurélie Marchet, Mathieu Médina, Sylvain Moineau, Christophe Novou-dit-Picot, Jean-Paul Pirnay, Fabrice Pirot, Grégory Resch, Forest Rohwer, Denise Tremblay, Gilbert Verbeken, and those who wish to remain anonymous or whom I have chosen to reference under a pseudonym. I also thank those employed by the French National Agency for the Safety of Medicines and Health Products for their warm welcome and availability, particularly Caroline Semaille and Nathalie Morgenzstejn.

The seminar on multiscalar ethnographies organized by Emilia Sanabria, in the beautiful presence of Kris Peterson and Valerie Olson, with the participation of Claire Beaudevin, Fanny Chabrol, Denis Chartier, Joe Dumit, Sophie Houdart, Silvia Mesturini, Mariana Rios, and Piera Talin, was the real starting point for this work. It was on this occasion that Sophie Houdart agreed to accompany me in my habilitation to supervise research—for which this book was written—discussed and generously commented on by Soraya Boudia, Maurice Cassier, Grégory Delaplace, Bruno Latour, Frédéric Le Marcis, Emilia Sanabria, and Isabelle Stengers. My sincere thanks to them.

Some concepts developed in this work were discussed in the Arctic Circle with members of the Kilpisjärvi Collective: Sally Atkinson, Sabine Biedermann, Andrea Butcher, Jose Canada, Denis Chartier, A. C. Davidson, Mark Erickson, Joshua Evans, Nicolas Fortané, Veera Kinnunen, Marine Legrand, Elina Oinas, Matthäus Rest, Salla Sariola, Andie Thompson, and Catherine Will.

I have also benefited from feedback and advice from Samuel Alizon, Henri Boulier, Clare Chandler, Nathalie Chazal, Laurie Denyer Willis, John Dupré, Mathilde Gallay-Keller, Lucie Gerber, Sophie Gerber, Patrick Giraudoux, Andrea-Luz Guttierez-Choquevilca, Maël Lemoine, Arnaud Macé, Germain

Meulemans, Jean-François Moreau, Gilles Moutot, Fabienne Orsi, Thomas Pradeu, Gaëlle Ronsin, Mélanie Roustan, Jérôme Santolini, Thibault Serviant-Fine, Charles Stépanoff, Anna Tsing, Paul Turner, and Frédéric Vagneron.

This research was made possible thanks to funding from the French National Research Agency and the Nouvelle-Aquitaine Region (and the unpaid work of evaluators), as well as those who accompanied me on these projects: Tristan Berger, Thomas Bonnin, and Jessica Pourraz, with whom I had enriching and fruitful discussions and from whom I have learned so much; Brigitte Pailley, whose meticulous management work is indispensable for conducting research; and the Centre Émile-Durkheim.

Over the years, I formed strong and rich friendships that nourished this book, friendships that sometimes emerge between the lines or in references that do not fully reflect what these people have contributed to me. My heartfelt thanks to Bruno Latour for offering me his friendship and for his thoughtfulness, his kindness, and his generosity in a world, that of research, which is increasingly lacking in these qualities. Bruno, I miss your presence and our discussions. Thanks to Kenza Afsahi, Sarah Al-Matary, Marine Boutin, Émilie Hache, Emilia Sanabria, and Alina Surubaru: Beyond their contributions to the ideas developed in this book, they remind me daily of the joy and power of collectives and give me strength and courage to overcome difficulties. Thanks to Alexis Zimmer and François Thoreau for our fertile and rhizomatic discussions, the accuracy of their advice, and the generosity of their readings.

I thoroughly enjoyed working with Victoria Denys, who created the beautiful illustrations that accompany and enrich the text, as well as with the French edition team at Amsterdam, who have done an outstanding editorial job: Lambert Clet, Eva Coly, and Lucien Perrin. I am especially grateful to Nicolas

Vieillescazes for encouraging me and then suggesting publishing my research (long before the publishing world became interested in viruses and microbes in general) and for his patient and meticulous proofreading. I am deeply thankful to Vincent Lépinay for offering me the opportunity to make this book exist in its English edition and for his presence, encouragement, and humor, and to Eric Schwartz for believing in this project. Thanks to Columbia University Press, particularly Alyssa Napier and Jennifer Crewe.

My curiosity for quirky topics since a young age has always been nurtured and encouraged by my parents, Pascale and Gérard Brives: Without their love and support, I would not have had the chance to continue my explorations into adulthood. Thanks to my sisters, Anne-Laure Brives and Agathe Lecacheur-Brives, for their support and encouragement and for all the moments, beautiful, joyful, sad, and painful, that we go through together.

This book was written with my children in mind (and often in the room), Edgar and Adèle—small pluribiotic beings, inexhaustible sources of joy, wonder, and transformation, and who have since been joined by Félix—, thinking of their futures, as well as those of Emmett, Lisbeth, Meryl, and Simone. This book would not be the same without Julien Théry: The rigor and integrity of his thought, his way of being in the world, his curiosity, his love, and his patient listening made the writing both softer and more intense. I thank him for sharing this experience (and so many others, and it's not over!) with me.

NOTES

INTRODUCTION

1. World Health Organization, *Global Action Plan on Antimicrobial Resistance* (World Health Organization, 2014), ix.
2. L'Assurance Maladie, "Antibiorésistance: comment mieux utiliser les antibiotiques?," January 4, 2022, https://www.ameli.fr/assure/sante/medicaments/comprendre-les-differents-medicaments/antibioresistance.
3. Antimicrobial Resistance Collaborators, "Global Burden of Bacterial Antimicrobial Resistance in 2019: A Systematic Analysis," *Lancet* 399, no. 10325 (2022): 629–55.
4. See, in particular, Andrew Singer et al., "Reinventing the Antimicrobial Pipeline in Response to the Global Crisis of Antimicrobial-Resistant Infections," *F1000 Research* 8 (2019): 238. In this article, the authors call for reflection on how to reinvent antibiotic production and industry supply chains.
5. For a summary of these alternatives, see, for example, Lloyd Czaplewski et al., "Alternatives to Antibiotics—a Pipeline Portfolio Review," *Lancet Infectious Diseases* 16, no. 2 (2016): 239–51.
6. Steffanie Strathdee and Thomas Patterson, *The Perfect Predator: A Scientist's Race to Save Her Husband from a Deadly Superbug* (Hachette, 2019). In this work, Strathdee, a biologist at the University of San Diego, California, describes how she saved the life of her husband, who had been infected with an aggressive, antibiotic-resistant pathogenic

bacterium during a trip to Egypt, thanks to help from her colleagues, a powerful scientific network, and phages.

7. Bacteriophage viruses are put to more complex uses in the treatment of bacterial infections, which I will not be exploring in the remainder of this book. See, for example, Benjamin Chan et al., "Phage Selection Restores Antibiotic Sensitivity in MDR *Pseudomonas aeruginosa,*" *Scientific Reports* 6, no. 26717 (2016).

8. This is a reference to the marvelous book by Lynn Margulis and Dorion Sagan, *Microcosmos: Four Billion Years of Microbial Evolution* (University of California Press, 1997).

9. Gilles Deleuze and Félix Guattari, *A Thousand Plateaus: Capitalism and Schizophrenia,* trans. Brian Massumi (University of Minnesota Press, 1987), 3–4. Originally published as *Mille plateaux: capitalisme et schizophrenie 2* [Minuit, 1980.]).

10. A large proportion of microorganisms cannot be cultivated in the laboratory, notably because they need the presence of other microorganisms to survive. See, in particular, Maureen O'Malley, *Philosophy of Microbiology* (Cambridge University Press, 2014), which provides a detailed analysis of the questions raised by the development of knowledge about microorganisms.

11. In this book, I steer clear of the endless controversies about whether viruses should be considered part of the living world. The main reason for excluding them is their inability to reproduce without their host, which is why, as mentioned earlier, they are called "obligate" parasites. This conception of obligation is becoming increasingly inoperative as new biological knowledge reveals that all living things are more dependent—at all levels—on other forms of life than previously thought. (I explore this relational ontology in detail in chapter 4). Gladys Kostyrka, "La place des virus dans le monde vivant" (PhD diss., Université Paris 1 Panthéon Sorbonne, 2018) provides a comprehensive and precise overview of these controversies since the sixteenth century. For further details about the influence of viruses on the philosophy of biology, I also recommend John Dupré and Maureen O'Malley, "Varieties of Living Things: Life at the Intersection of Lineage and Metabolism," *Philosophy & Theory in Biology* 1, no. 3 (2009); and the special issue on viruses coordinated by Thomas Pradeu et al., "Understanding Viruses:

Philosophical Investigations," *Studies in History and Philosophy of Biological and Biomedical Sciences* 59 (2016): 57–63.

12. The field of multispecies studies encompasses various approaches, ranging from ethnography to geography, and is characterized by a strong focus on the ethical, political, and epistemological issues associated with the recognition of the relationships (or lack thereof) between humans and other forms of life. See Thom van Dooren et al., eds., "Multispecies Studies: Cultivating Arts of Attentiveness," special issue, *Environmental Humanities* 8, no. 1 (2016).

13. Heather Paxson and Stefan Helmreich, "The Perils and Promises of Microbial Abundance: Novel Natures and Model Ecosystems, from Artisanal Cheese to Alien Seas", *Social Studies of Science* 44, no. 2 (2014): 165–93. The first to have theorized and demonstrated this notion is Bruno Latour, in his book devoted to Louis Pasteur, *The Pasteurization of France* (Harvard University Press, 1988). For more recent work, see Stefan Helmreich, *Alien Ocean: Anthropological Voyages in Microbial Seas* (University of California Press, 2009); Heather Paxson, *The Life of Cheese: Crafting Food and Value in America* (University of California Press, 2012); Amber Benezra, "Race in the Microbiome," *Science, Technology & Human Values* 45, no. 5 (2020): 877–902; Beth Greenhough et al., "Unsettling Antibiosis: How Might Interdisciplinary Researchers Generate a Feeling for the Microbiome and to What Effect?," *Palgrave Communications* 4, no. 149 (2018); Charlotte Brives et al., eds., *With Microbes* (Mattering, 2021). For a critical overview of studies on microbes in the social sciences, see Charlotte Brives and Alexis Zimmer, eds., "Ecologies and Promises of the Microbial Turn," special issue, *Revue d'anthropologie des connaissances* 15, no. 3 (2021).

14. These conditions are referred to as examples of "dysbiosis." For a critical analysis of the concept of dysbiosis and its use, see Katarzyna B. Hooks and Maureen A. O'Malley, "Dysbiosis and Its Discontents," *mBio* 8, no. 5 (2017): e01492–17.

15. Jamie Lorimer, *The Probiotic Planet: Using Life to Manage Life* (University of Minnesota Press, 2020), 2.

16. Emily Martin was the first to analyze martial rhetoric, particularly in her work on immunity, *Flexible Bodies* (Beacon, 1995). Regarding the use of phages during the COVID-19 pandemic, see Charlotte Brives, "The

Politics of Amphibiosis: The War Against Viruses Will Not Take Place," *Somatosphere: Science, Medicine, and Anthropology*, April 19, 2020, https:// shs.hal.science/halshs-02873882/file/the-politics-of-amphibiosis .pdf; Charlotte Brives, "Pluribiose. Vivre avec les virus. Mais comment?," Terrestres 14 (2020); Bernadette Bensaude-Vincent, "Guerre et paix avec le coronavirus," Terrestres 13 (2020); and Lorenzo Servitje, *Medicine Is War: The Martial Metaphor in Victorian Literature and Culture* (State University of New York Press, 2021).

17. Multispecies studies pay particular attention to these subtle interrelationships, whose descriptions are now more indispensable than ever, to understand the effects of human activities on other species that are likely to have a significant interaction with humans (i.e., to "talk" to them). See, in particular, Thom van Dooren, *Flight Ways: Life and Loss at the Edge of Extinction* (Columbia University Press, 2014).

18. Regarding virus domestication and viruses as companion species, see Charlotte Brives, "From Fighting Against to Becoming With: Viruses as Companion Species," in *Hybrid Communities: Biosocial Approaches to Domestication and Other Trans-species Relationships*, ed. C. Stépanoff and J.-D. Vigne (Routledge, 2017), 115–26; and Beth Greenhough, "Where Species Meet and Mingle: Endemic Human–Virus Relations, Embodied Communication and More-Than-Human Agency at the Common Cold Unit 1946–90," *Cultural Geographies* 19, no. 3 (2012): 281–301.

19. The importance of prioritizing certain relationships over others is particularly apparent when it comes to putting microbes at the service of humans, especially in fermentation practices with their complex interplay of multiple equilibria. See Jessica Hendy et al., eds., "Cultures of Fermentation," *Current Anthropology* 62, no. S24 (October 2021). For example, Heather Paxson coined the concept of "microbiopolitics" to account for these complex adjustments. See Heather Paxson, "Post-Pasteurian Cultures: The Microbiopolitics of Raw-Milk Cheese in the United States," *Cultural Anthropology* 23, no. 1 (2008): 15–47.

20. Isabelle Stengers, *In Catastrophic Times: Resisting the Coming Barbarism* (Open Humanities, 2015).

21. I underline in passing, and in agreement with the work of Soraya Boudia and Nathalie Jas, Gouverner un monde toxique (Quae, 2019), that when it comes to establishing scientific knowledge about environmental

disasters linked to chemical pollution, there is no equivalent to the entity addressing climate change: the Intergovernmental Panel on Climate Change, a body of the United Nations. I should add that antibiotics are never considered on the same level as other chemical molecules, even though—as demonstrated by antibiotic resistance—they can also generate pollution.

22. Bruno Latour, "Give Me a Laboratory and I Will Raise the World," in *Science Observed: Perspectives on the Social Study of Science*, ed. K. D. Knorr-Cetina and M. Mulkay (Sage, 1983), 155.

23. On the concept of biocapital, see, in particular, Kaushik Sunder Rajan, *Biocapital* (Duke University Press, 2006).

24. Melinda Cooper, *Life as Surplus: Biotechnology and Capitalism in the Neoliberal Era* (University of Washington Press, 2008).

25. The notion of situated knowledge was forged by Donna Haraway to create a space covering possibilities between two unsatisfactory positions: the position represented by advocates of a disembodied scientific objectivity (the "view from nowhere") and the position represented by relativistic philosophies, according to which all constructs are equal because they are constructs. See Donna Haraway, "Situated Knowledges: The Science Question in Feminism and the Privilege of Partial Perspective," *Feminist Studies* 14, no. 3 (1988): 575–99.

26. This is a position defended in Donna Haraway, *Modest_Witness@Second _Millenium. FemaleMan_Meets_OncoMouse: Feminism and Technoscience* (Routledge, 1997). On the noninnocence of researchers, see also Vinciane Despret, "En finir avec l'innocence, dialogue avec Isabelle Stengers et Donna Haraway," in *Penser avec Donna Haraway*, ed. E. Dorlin and E. Rodriguez (Puf, 2012), 23–45. On alternative figures of the scientist and other ways of practicing science, see Benedikte Zitouni, "Héritières de la Révolution scientifique: d'autres figures et manières de faire science," *Nouvelles Questions Féministes* 40, no. 2 (2021): 35–51.

27. On this point, see the article by François Thoreau concerning the "embedment" of social scientists and the research policies that such embedment should bring to light. François Thoreau, "L'embarquement par son objet, trois politiques de l'enquête sur les clôtures virtuelles (*virtual fences*)," *Revue d'anthropologie des connaissances* 13, no. 2 (2019): 399–423.

28. Marylin Strathern, *Partial Connections* (AltaMira, 1991). I thank Emilia Sanabria for reminding me of this connection when she read an earlier version of this text. Work on how microbiology affects the practice of anthropology—notably the complex question of totalities, "individuals," or "societies," for example—is currently in progress and therefore not covered in this book.

29. Kim Fortun, "Ethnography in Late Industrialism," *Cultural Anthropology* 27, no. 3 (2012): 460.

30. This limited focus calls for a considerable amount of further study, especially on the initiatives being developed in countries of the Global South, particularly on the African continent, which, as studies show, has been hit the hardest by bacterial resistance to antibiotics.

1. TENSIONS

1. I have used a pseudonym at the request of the main protagonist in this chapter. Pseudonyms are also used for the two doctors he mentions in his narrative, this time owing to the sensitivity of the subject and the strategies they adopted.

2. People with chronic illnesses, infectious or otherwise, often learn to interpret their symptoms as clues to their condition; for example, some people with diabetes can accurately predict their blood sugar levels without needing to measure them (Annemarie Mol, *The Logic of Care: Health and the Problem of Patient Choice* [Routledge, 2008]). Pierre (also a pseudonym), a man who has been suffering from multiple chronic infections for over forty years, said during an interview, "You know, I couldn't describe to you the kind of pain—a dull pain—that you feel when staph [i.e., *Staphylococcus* bacteria] is present, but it's recognizable, and I knew that my staph had returned."

3. With regard to "adopting the wrong approach," André used the French term "maladresse," which can be broken down into "mal-adresse" or "mal s'adresser à" and means "to speak inappropriately [to someone]." Here, André is referring to his desire to avoid treating his bacterial infections ineffectively.

4. For a detailed report of this meeting, see Agence nationale de sécurité du médicament (ANSM), *Compte rendu de séance: Comité scientifique*

spécialisé temporaire. Phagothérapie—retour d'expérience et perspectives (ANSM, 2019).

5. ANSM, *Compte rendu de séance*, 11.

6. See, for example, Jamie Lorimer, *The Probiotic Planet: Using Life to Manage Life* (University of Minnesota Press, 2020), which discusses other therapies involving biological entities such as helminths.

7. Abundant literature exists on the regulation of medicinal products. For a particularly illuminating historical analysis of the various types of regulations, logics, and actors involved, see Jean-Paul Gaudillière and Volker Hess, eds., *Ways of Regulating Drugs in the 19th and 20th Centuries* (Palgrave Macmillan, 2013).

2. ALTERNATIVE HISTORIES

1. Interview with André, May 2018.

2. Recounting this history will take some time as it involves describing both phage therapy practices and what made them possible at different times and in different places, as well as analyzing the links that the community has maintained with two other life science communities that have used phages. These two communities include fundamental research, notably with the development of molecular biology (which I will return to briefly in chapter 4), and applied research, with the use of phages to identify bacterial strains responsible for specific epidemics (a process called "phage typing"), particularly in the former French, British, and German colonies. See Claas Kirchhelle, "The Forgotten Typers," *Notes and Records* 74, no. 4 (2019): 539–65.

3. For further details, see, for example, Emiliano Fruciano, "La Phagothérapie. Émergence d'une idée controversée et logique d'un échec (1917–1949)" (PhD diss., École des hautes études en sciences sociales, 2011); and Anna Kuchment, *The Forgotten Cure: The Past and Future of Phage Therapy* (Copernicus, 2012). Sinclair Lewis gave phage therapy a starring role in his novel *Arrowsmith* (1925), whose main character, partly inspired by d'Hérelle but mainly Paul de Kruif (who collaborated with Lewis in the writing of the novel), uses bacteriophage viruses to save the population of St. Hubert, an imaginary island in the Caribbean, from a bubonic plague epidemic. The novel describes a fictional clinical

trial and Arrowsmith's desire to prove the efficacy of these entities scientifically and methodically, even though very few trials had been undertaken at that time. For an epistemological interpretation of *Arrowsmith*, see Ilana Löwy, "Martin Arrowsmith's Clinical Trial: Scientific Precision and Heroic Medicine," *Journal of the Royal Society of Medicine* 1 (2010): 461–66.

4. In 1923, for example, a trial using ten thousand ampoules of phage preparations to treat dysentery had positive results. And in 1924, a phage preparation called Bacteriofagina dysenterica was widely used by government troops during the revolution in São Paulo, Brazil, and was included in medical textbooks in the late 1920s. See Gabriel Magno de Freitas Almeida and Lotta-Riina Sundberg, "The Forgotten Tale of Brazilian Phage Therapy," *Lancet Infectious Diseases* 20, no. 5 (2020): e90–e101; Fruciano, "La phagothérapie"; and Thomas Häusler, *Viruses vs. Superbugs: A Solution to the Antibiotics Crisis?* (Macmillan, 2006).

5. Fruciano, "La phagothérapie"; Kuchment, *The Forgotten Cure*; Häusler, *Viruses vs. Superbugs*.

6. Häusler, *Viruses vs. Superbugs*; Fruciano, "La phagothérapie."

7. This is revealed by an examination of the protocols of clinical trials conducted during the interwar period, which gives rise to a major epistemic problem: how to provide evidence of the efficacy of a therapy when there is no consensus, at least on the mode of existence of what is being evaluated. Two overview articles on phage therapy, published in the prestigious *Journal of the American Medical Association* in 1934 and 1941, highlight these difficulties. See Monroe D. Eaton and Stanhope Bayne-Jones, "Bacteriophage Therapy," *Journal of the American Medical Association* 103 (1934): 1769–76; and Albert Paul Krueger and E. Jane Scribner, "The Bacteriophage," *Journal of the American Medical Association* 116 (1941): 2160–69.

8. Freddy Himmelweit, "Combined Action of Penicillin and Bacteriophage on *Staphylococci*," *Lancet* 246, no. 6361 (1945): 104–105.

9. Gaëlle Bourgeois, "Histoire de la phagothérapie à Lyon" (MD thesis, Université Claude-Bernard Lyon 1, 2020).

10. The *Vidal* is France's equivalent of America's *Prescribers' Digital Reference* and the United Kingdom's *Monthly Index of Medical Specialities*.

11. Since then, Alain Dublanchet has published two books containing a wealth of information: Alain Dublanchet, *La Phagothérapie. Des virus*

pour combattre les infections (Favre, 2017); and Alain Dublanchet, *Autobiographie de Félix d'Hérelle. Les pérégrinations d'un bactériologiste* (Éditions médicales internationales, 2017).

12. The personality of d'Hérelle, who appears to have been an eccentric, controversial, and uncompromising scientist, has been the subject of several publications and biographies. See Dublanchet, *Autobiographie de Félix d'Hérelle*; William C. Summers, "On the Origins of the Science in *Arrowsmith*: Paul de Kruif, Félix d'Hérelle, and Phage," *Journal of the History and Medicine and Allied Sciences* 46 (1991): 315–32; and Fruciano, "La phagothérapie."

13. Dmitry Myelnikov, "An Alternative Cure: The Adoption and Survival of Bacteriophage Therapy in the USSR, 1922–1955," *Journal of the History of Medicine and Allied Sciences* 73, no. 4 (2018): 385–411. The implementation of the mass production of penicillin was greatly facilitated by the decision taken at the Tehran Conference, convened by Winston Churchill, Joseph Stalin, and Franklin Delano Roosevelt in 1943 in preparation for the Normandy landings of the Second World War, to send four American and British scientists to Moscow to exchange scientific knowledge, including in relation to the production of penicillin.

14. For further insight into the ecological conception of diseases, see Warwick Anderson, "Natural Histories of Infectious Disease: Ecological Vision in Twentieth-Century Biomedical Science," *Osiris* 19 (2004): 39–61.

15. The role of phages in the Battle of Stalingrad is probably the most common anecdote that I have heard in the field in recent years. In the besieged city, the microbiologist Zinaida Yermolyeva produced and distributed large quantities of an anti-cholera phage preparation she had developed three years earlier, thereby preventing the outbreak of an epidemic in the ranks of the Red Army.

16. Myelnikov, "An Alternative Cure," 4.

17. I have used quotation marks here to emphasize my interlocutors' use of a general, undifferentiated category when in fact Tbilisi is home to several phage treatment centers—Eliava being only the oldest and best known of them—and therefore also likely home to a diverse range of techniques and practices.

18. Interview with Alain-Michel Ceretti, March 2018.

19. "Medicinal Product," European Medicines Agency, accessed November 20, 2021, https://www.ema.europa.eu/en/glossary-terms/medicinal -product. This definition is based on *Directive 2001/83/EC of the European Parliament and of the Council of 6 November 2001 on the Community Code Relating to Medicinal Products for Human Use*, available at https:// eur-lex.europa.eu/eli/dir/2001/83/oj/eng.

20. "Early access authorization" is governed by Article 83 of Regulation (EC) No. 726/2004 of March 31, 2004, laying out the European Community procedures for the authorization and supervision of medicinal products; Article 78 of Law No. 2020–1576 of December 14, 2020, on the financing of social security for 2021; and Article L. 5121-12-1 of the French Public Health Code.

21. Ethical liability on the basis of Articles R.4127-8, R.4127-32, R.4127-39, and R.4127-40 of the French Public Health Code; civil liability on the basis of Articles L.5121-12-1, L.1110-5, and L.1142-1 I of the same code; and criminal liability on the basis of Articles 223-1 and 222-19 of the French Criminal Code.

22. These products are usually provided by the start-up Pherecydes or a team from the Queen Astrid Military Hospital in Brussels, which can obtain them from a research laboratory if it does not possess any active phages for the bacteria of interest. I will return to these aspects in subsequent chapters.

23. Kuchment, *The Forgotten Cure*, 67–69.

24. For an illuminating historical analysis of drug regulation recounting the various approaches and actors involved, see Jean-Paul Gaudillière and Volker Hess, eds., *Ways of Regulating Drugs in the 19th and 20th Centuries* (Palgrave Macmillan, 2013).

3. MICROGEOHISTORIES

1. Jacques Pépin, *The Origins of AIDS* (Cambridge University Press, 2011).
2. Jörg Niewöhner and Margaret Lock, "Situating Local Biologies: Anthropological Perspectives on Environment/Human Entanglements," *BioSocieties* 13 (2018): 681–97; Anna Tsing, *The Mushroom at the End of the World: On the Possibility of Life in Capitalist Ruins* (Princeton University Press, 2015).

3. Lederberg also played an important role in the production of knowledge associated exclusively with her husband, Joshua Lederberg, who did not even mention her in his Nobel Prize acceptance speech. I would like to thank Frédéric Vagneron for pointing out her absence in an earlier version of this text, yet another insult to the invisibilized work of minorities. For a biography of Esther Lederberg, see Thomas E. Schindler, *A Hidden Legacy: The Life and Work of Esther Zimmer Lederberg* (Oxford University Press, 2021), which includes a chapter on the Matilda effect (the systemic denial, despoliation, or minimization of women's contribution to scientific research, whose work is often attributed to their male colleagues), illustrated by the story of the invisibilization of three other women: Martha Chase, Laura Garnjobst, and Daisy Dussoix. For a history of molecular biology that puts the spotlight on the people and groups marginalized by the history of "great men," see Mathias Grote et al., "The Molecular Vista: Current Perspectives on Molecules and Life in the Twentieth Century," *History and Philosophy of the Life Sciences* 43, no. 1 (2021), which describes how this type of invisibilization is linked to an invisibilization of knowledge and know-how that are also considered negligible or inferior.

4. For further details, see Lily Kay, *The Molecular Vision of Life* (Oxford University Press, 1993); Michel Morange, *Histoire de la biologie moléculaire* (La Découverte, 1994); Mireille Ansaldi et al., "Un siècle de recherche sur les bacteriophages," *Virologie* 24, no. 1 (2020): 9–22; and Mireille Ansaldi et al., "Les applications antibactériennes des bacteriophages," *Virologie* 24, no. 1 (2020): 23–36.

5. I insist on the following points. Although phages can be found everywhere, this does not mean that they can be used as is. The purification of phage preparations is a technical challenge since it is necessary to ensure that nothing in the preparation can become hazardous when the phages are administered. They are certainly present in wastewater, but this does not mean that patients can be cured by using foul water.

6. Grégory Resch has since joined the Centre hospitalier universitaire vaudois in Lausanne, where he continues to work on the characterization of relationships between phages and bacteria in the context of phage therapy development.

7. These bacteria are grouped together under the acronym ESKAPE (*Escherichia coli*, *Staphylococcus aureus*, *Klebsiella pneumoniae*, *Acineto-bacter baumannii*, *Pseudomonas aeruginosa*, and *Enterobacter sp.*).

8. For further information about collections as an archive of the history of a laboratory, see Charlotte Brives, "Des levures et des hommes. Anthropologie des relations entre humains et non humains dans un laboratoire de biologie" (PhD diss., Université Victor Segalen, 2010).

9. I have used a pseudonym at her request, but I would like to thank Julie once again for her warm welcome.

10. However, she does more than simply put each bacterium in proximity to each phage. For each crossing, she also varies a parameter: the quantity of phages. Each phage is initially collected from a liquid stock solution, usually highly concentrated, with an approximate titer of 10^9, corresponding to around one billion phages per milliliter. If the phage is active on the bacteria, it will be impossible to count the lysis plaques for such a concentration because the entire bacterial mat will be destroyed. Julie therefore dilutes her initial solution by a factor of ten on each line to ensure that one of the eight dilutions will enable her to count the lysis plaques and therefore "titrate" her initial solution.

11. For information about ephemeral inscriptions and chains of inscriptions, see, for example, Bruno Latour and Steve Woolgar, *Laboratory Life: The Social Construction of Scientific Facts* (Sage, 1979); and Charlotte Brives, "Le rôle des écrits éphémères dans la production des faits scientifiques: la domestication de la levure sauvage," Langage et Société 127 (2009): 71–81.

12. See, for example, Brives, "Des levures et des hommes." See also Vinciane Despret's research on ethology, notably *La Danse du cratérope écaillé. Naissance d'une théorie éthologique* (Les Empêcheurs de Penser en Rond, 1996), which describes studies that have inspired me ever since working on my thesis and which have taught me about the "politeness of getting to know each other."

13. At that time (in 2019), this phage was then sent to the team at Queen Astrid Military Hospital in Brussels for production and preparation to enable it to be administered to the sick person. I will revisit this episode in chapter 9.

14. Culture media are essential in the life sciences, especially since they are involved in the production and stabilization of biological entities.

See Hannah Landecker, *Culturing Life: How Cells Became Technologies* (Harvard University Press, 2006); and Charlotte Brives, "Des levures et des hommes."

15. In *Facing Gaia: Eight Lectures on the New Climatic Regime* (Polity, 2017), Bruno Latour uses the term *gaiahistory*, which he uses in opposition to *history*. Donna Haraway takes up the distinction between geohistory and the history of human control over nature in *Staying with the Trouble: Making Kin in the Chthulucene* (Duke University Press, 2016).

16. Robert Kohler, *Lords of the Fly: Drosophila Genetics and the Experimental Life* (University of Chicago Press, 1994), 6.

17. On model organisms in biology, see Bonnie Clause, "The Wistar Rat as a Right Choice: Establishing Mammalian Standards and the Ideal of a Standardized Mammal," *Journal of the History of Biology* 26, no. 2 (1993): 329–49; Rachel Ankeny, "The Conqueror Worm: An Historical and Philosophical Examination of the Use of the Nematode *Caenorhabditis Elegans* as a Model Organism" (PhD diss., University of Pittsburgh, 1997); Angela Creager, *The Life of a Virus: Tobacco Mosaic Virus as an Experimental Model, 1930–1965* (University of Chicago Press, 2002); Karen Rader, *Making Mice: Standardizing Animals for American Biomedical Research, 1900–1955* (Princeton University Press, 2004); Sophie Houdart, *La Cour des miracles. Ethnologie d'un laboratoire japonais* (CNRS, 2008); Kohler, *Lords of the Fly*; and Brives, "Des levures et des hommes."

18. On the role of cryogenics in the life sciences and how it enables the manipulation of temporalities, see Landecker, *Culturing Life*. See also Joanna Radin and Emma Kowal, eds., *Cryopolitics: Frozen Life in a Melting World* (MIT Press, 2017).

19. In "Des levures et des hommes," a laboratory ethnography conducted on the relationships between humans and *Saccharomyces cerevisiae* yeast, I described how access to the -80°C freezer containing the laboratory's yeast collection was strictly regulated. The comings and goings around the freezer revealed how the laboratory was organized, the implicit and explicit norms, and the rules put in place to facilitate the yeast specialists' coexistence. These measures were supposed to ensure that the yeast strains in the freezer remained identical in all respects and could continue to act as "reliable witnesses" to experimentation and continue to "respond to the response," to use an expression coined by Isabelle

Stengers in *The Invention of Modern Science* (University of Minnesota Press, 2000).

20. Brives, "Des levures et des hommes."

21. The notion of recalcitrance is developed in Adele Clarke and Joan Fujimura, eds., *The Right Tools for the Job: At Work in Twentieth-Century Life Sciences* (Princeton University Press, 1992).

4. PLURIBIOSIS

The quote in the epigraph to this chapter comes from Forest Rohwer et al., *Life in Our Phage World: A Centennial Field Guide to the Earth's Most Diverse Inhabitants* (Wholon, 2014), xv.

1. Mireille Ansaldi et al., "Un siècle de recherche sur les bacteriophages," *Virologie* 24, no. 1 (2020): 16.

2. Lynn Sagan, "On the Origin of Mitosing Cells," *Journal of Theoretical Biology* 14, no. 3 (1967): 225–74. For a longer version brimming with examples, see Lynn Margulis and Dorion Sagan, *Microcosmos: Four Billion Years of Microbial Evolution* (University of California Press, 1997).

3. Scott Gilbert et al., "A Symbiotic View of Life: We Have Never Been Individuals," *Quarterly Review of Biology* 87, no. 4 (2012): 325–34. For information about how the social sciences are taking up these questions, see Amber Benezra et al., "Anthropology of Microbes," *Proceedings of the National Academy of Sciences* 109, no. 17 (2012): 6378–81; Funke Iyabo Sangodeyi, "The Making of the Microbial Body, 1900s–2012" (PhD diss., Harvard University, 2014); Stefan Helmreich, "Homo Microbis: The Human Microbiome, Figural, Literal, Political," *Threshold* 42 (2014): 52–59; Heather Paxson and Stefan Helmreich, "The Perils and Promises of Microbial Abundance: Novel Natures and Model Ecosystems, from Artisanal Cheese to Alien Seas," *Social Studies of Science* 44, no. 2 (2015): 165–93; Tobias Rees et al., "How the Microbiome Challenges Our Concept of Self," *PLoS Biology* 16, no. 2 (2018): e2005358; Alexis Zimmer, "Collecter, conserver, cultiver des microbiotes intestinaux. Une biologie du sauvetage," *Écologie & Politique* 58 (2019): 135–50; Beth Greenhough et al., "Unsettling Antibiosis: How Might Interdisciplinary Researchers Generate a Feeling for the Microbiome and to What Effect?," *Palgrave Communications* 4, no. 149 (2018);

and Amber Benezra, "Race in the Microbiome," *Science, Technology & Human Values* 45, no. 5 (2020): 877–902.

4. Éric Bapteste, *Tous entrelacés! Des gènes aux super-organismes: les réseaux de l'évolution* (Belin, 2017); Ed Yong, *I Contain Multitudes: The Microbes Within Us and a Grander View of Life* (HarperCollins, 2016); Marc-André Selosse, *Jamais seul. Ces microbes qui construisent les plantes, les animaux et les civilisations* (Actes Sud, 2017).

5. Merry Youle, *Thinking Like a Phage: The Genius of the Viruses That Infect Bacteria and Archaea* (Wholon, 2017), 18.

6. Steven Wilhelm and Curtis Suttle, "Viruses and Nutrient Cycles in the Sea," *Bioscience* 49, no. 10 (1999): 781–88.

7. Curtis Suttle, "Marine Viruses—Major Players in the Global Ecosystem," *Nature Reviews* 5, no. 10 (2007): 804.

8. Peter Peduzzi et al., "The Virus's Tooth: Cyanophages Affect an African Flamingo Population in a Bottom-Up Cascade," *ISME Journal* 8, no. 6 (2014): 1346–51.

9. Harald Brüssow et al., "Phages and the Evolution of Bacterial Pathogens: From Genomic Rearrangements to Lysogenic Conversion," *Microbiology and Molecular Biology Reviews* 68, no. 3 (2004): 560–602.

10. Graham Hatfull, "Dark Matter of the Biosphere: The Amazing World of Bacteriophage Diversity," *Journal of Virology* 89, no. 16 (2015).

11. Another variant is based on the "red queen" hypothesis. According to this hypothesis, a reference to *Alice's Adventures in Wonderland*, the current biodiversity is the result of interactions between organisms that led to endless evolutionary races. For details about these many coevolutionary mechanisms, see, for example, Alex Betts et al., "Contrasted Coevolutionary Dynamics Between a Bacterial Pathogen and Its Bacteriophages," *Proceedings of the National Academy of Sciences* 111, no. 30 (2014).

12. Rohwer et al., *Life in Our Phage World*, 2–43.

13. The historian of science Evelyn Fox Keller has devoted part of her career to analyzing the role of metaphors in the life sciences and their changes. See Evelyn Fox Keller, *The Century of the Gene* (Harvard University Press, 2002); and Evelyn Fox Keller, *Making Sense of Life: Explaining Biological Development with Models, Metaphors, and Machines* (Harvard University Press, 2002). In a masterful paper, Emily

Martin has demonstrated the role of gender stereotypes in the production of knowledge in biology, particularly in reproductive biology. See Emily Martin, "The Egg and the Sperm: How Science Has Constructed a Romance Based on Stereotypical Male–Female Roles," *Journal of Women in Culture and Society* 16, no. 3 (1991): 485–501. On the use of warlike metaphors in the conception of immunity, see Emily Martin, *Flexible Bodies* (Beacon, 1995). See also Eben Kirksey, "Queer Love, Gender Bending Bacteria, and Life After the Anthropocene," *Theory, Culture & Society* 36, no. 6 (2019): 197–219.

14. Priscilla Wald, *Contagious: Cultures, Carriers, and the Outbreak Narrative* (Duke University Press, 2008). While Wald provides a literary analysis of this issue, a great deal of research has shown, in remarkable fashion, how the solutions implemented to manage epidemics reproduce racist and sexist stereotypes.

15. "Patient zero" is typically blamed for "deviant" behavior and questionable morals. Such is the story of the stigmatization of Gaëtan Dugas, a young gay flight attendant who was long blamed for the HIV epidemic in North America. Steven Soderbergh's 2011 film *Contagion* also falls into this trap by making Gwyneth Paltrow's character, an adulterer, responsible for the spread of a virus worldwide. For an analysis of warlike rhetoric in COVID-19 pandemic narratives and their consequences for the political management of the health crisis, see, for example, Charlotte Brives, "The Politics of Amphibiosis: The War Against Viruses Will Not Take Place," *Somatosphere: Science, Medicine, and Anthropology*, April 19, 2020; and Bernadette Bensaude-Vincent, "Guerre et paix avec le coronavirus," *Terrestres* 13 (2020). On warlike metaphors in medicine, see Lorenzo Servitje, *Medicine Is War: The Martial Metaphor in Victorian Literature and Culture* (State University of New York Press, 2021).

16. See, for example, Carolyn Merchant, *The Death of Nature: Women, Ecology, and the Scientific Revolution* (HarperCollins, 1990).

17. For more details, See Anne Dupressoir and Thierry Heidmann, "Les syncytines, des protéines d'enveloppe rétrovirales capturées au profit du développement placentaire," *Médecine/Sciences* 27, no. 2 (2011): 163–69.

18. The analogy of gathering is not completely satisfactory, however, and will be the subject of further work.

19. Theodor Rosebury, *Microorganisms Indigenous to Man* (McGraw-Hill, 1962).

20. Shifting from describing the intrinsic properties of entities to describing the relationships established *between* these entities is the proposition I make in "The Politics of Amphibiosis."

21. This problem is found in also other concepts taken up by researchers in the human and social sciences studying microbes and, more broadly, relationships with nonhuman living beings. As I stated in the introduction, the notion of "probiotics," developed particularly in the work of Jamie Lorimer, implicitly maintains an opposition to the notion of "antibiotics," despite the author's intention being quite different. For a discussion of these concepts, see Charlotte Brives and Alexis Zimmer, "Ecologies and Promises of the Microbial Turn," *Revue d'anthropologie des connaissances* 15, no. 3 (2021).

22. This conception of living entities and their relations will not be fundamentally new to anyone familiar with certain schools of thought in the philosophy of biology. The virtually omnipresent notion of symbiosis—an intimate, long-lasting, and mutually beneficial association between two living organisms—had already undermined the notion of distinct entities. This conception has also led philosophers like John Dupré to develop a processual conception of life, that is, one based on dynamics and processes rather than things or substances. See John Dupré, *Processes of Life: Essays in the Philosophy of Biology* (Oxford University Press, 2012). In a 2016 paper, John Dupré and Stephan Guttinger demonstrate that viruses provide valuable examples for understanding the fundamentally interconnected and collaborative nature of nature: "Viruses as Living Processes," *Studies in History and Philosophy of Biological and Biomedical Sciences* 59 (2016): 109–16.

23. One could even say that to a certain extent, it eliminates the need to use such dichotomies if they are not explicitly used by those working in the field, which is why I refrain from discussing the nature–culture duality in this book (other than briefly in chapter 7).

24. Science and technology studies show us that there are various ways of addressing a research topic, various ways of framing it, bringing it to life, and enacting it. For an excellent example, see Annemarie Mol, *The Body Multiple: Ontology in Medical Practice* (Duke University Press, 2002).

25. Although the concept of pluribiosis originated from a description of research on living entities, it is also consistent with studies that go

beyond the biotic–abiotic dichotomy. This brings to mind the marvelous concept of intra-action formulated by Karen Barad in her work on physics: "Since individually determinate entities do not exist, measurements do not entail an interaction between separate entities; rather, determinate entities emerge from their intra-action. I introduce the term 'intra-action' in recognition of their ontological inseparability, as opposed to the usual 'interaction' which relies on a metaphysics of individualism (in particular, the prior existence of separately determinate entities). A phenomenon is a specific intra-action of an 'object' and the 'measuring agencies'; the object and the measuring agencies emerge from, rather than precede, the intra-action that produces them." Karen Barad, *Meeting the Universe Halfway: Quantum Physics and the Entanglement of Matter and Meaning* (Duke University Press, 2007), 128.

5. PLURIBIOTIC MEDICINE

Interview with Raphaëlle Delattre, November 27, 2018.

1. Paul Hyman and Stephen Abedon, "Bacteriophage Host Range and Bacterial Resistance," *Advances in Applied Microbiology* 70 (2010): 217–48.
2. Gerard D. Wright and Arlene D. Sutherland, "New Strategies for Combating Multidrug-Resistant Bacteria," *Trends in Molecular Medicine* 13, no. 6 (2007): 260–67.
3. Lucy Furfaro et al., "Bacteriophage Therapy: Clinical Trials and Regulatory Hurdles," *Frontiers in Cellular and Infection Microbiology* 8, no. 376 (2018); Patrick Jault et al., "Efficacy and Tolerability of a Cocktail of Bacteriophages to Treat Burn Wounds Infected by *Pseudomonas Aeruginosa* (PhagoBurn): A Randomised, Controlled, Double-Blind Phase 1/2 Trial," *Lancet Infectious Diseases* 19, no. 1 (2019): 35–45; Lorenz Leitner et al., "Intravesical Bacteriophages for Treating Urinary Tract Infections in Patients Undergoing Transurethral Resection of the Prostate: A Randomised, Placebo-Controlled, Double-Blind Clinical Trial," *Lancet Infectious Diseases* 21, no. 3 (2020): 427–36. Studies of animal models reach similar conclusions; see Nicolas Dufour et al., "Phage Therapy of Pneumonia Is Not Associated with an Overstimulation of the

Inflammatory Response Compared to Antibiotic Treatment in Mice," *Antimicrobial Agents and Chemotherapy* 63, no. 8 (2019): e00379–19; and Nicolas Dufour et al., "The Lysis of Pathogenic *Escherichia coli* by Bacteriophages Releases Less Endotoxin Than by β-Lactams," *Clinical Infectious Diseases* 64, no. 11 (2017): 1582–88. Because phage therapy is an experimental practice, each administration is systematically accompanied by an analysis of the phages' potential toxicity. The question of toxicity has therefore not yet been completely settled.

4. Interview with Tristan Ferry, April 19, 2019.

5. Interview with Raphaëlle Delattre, November 27, 2018.

6. In 2019, 1,700,000 deaths worldwide were caused by lower respiratory tract infections, 400,000 of which were directly attributable to antibiotic-resistant bacteria. See Antimicrobial Resistance Collaborators, "Global Burden of Bacterial Antimicrobial Resistance in 2019: A Systematic Analysis," *Lancet* 399, no. 10325 (2022): 629–55.

7. Warwick Anderson, "Natural Histories of Infectious Disease: Ecological Vision in Twentieth-Century Biomedical Science," *Osiris* 19 (2004): 39–61.

8. On other developments, see Amber Benezra et al., "Anthropology of Microbes," *Proceedings of the National Academy of Sciences* 109, no. 17 (2012): 6378–81; Stefan Helmreich, "Homo Microbis: The Human Microbiome, Figural, Literal, Political," *Threshold* 42 (2014): 52–59; and Charlotte Brives and Alexis Zimmer, "Ecologies and Promises of the Microbial Turn," *Revue d'anthropologie des connaissances* 15, no. 3 (2021).

9. Martin Blaser, *Missing Microbes: How the Overuse of Antibiotics Is Fueling Our Modern Plagues* (Picador, 2014). In this work, Blaser presents the example of the *Helicobacter pylori* bacterium, responsible for a significant proportion of stomach ulcers, which is now on the verge of extinction owing to the massive use of antibiotics, a trend correlated with an increase in esophageal cancers. Because controversy surrounds this example (which Blaser discusses extensively in his book), I have chosen to mention it in a footnote. For more on the notion of companion species, see Donna Haraway, *The Companion Species Manifesto* (University of Chicago Press, 2003). For an application of this concept to microbes, see Charlotte Brives, "From Fighting Against to Becoming With: Viruses as Companion Species," in *Hybrid Communities: Biosocial*

Approaches to Domestication and Other Trans-species Relationships, ed C. Stépanoff and J.-D. Vigne (Routledge, 2017); and Beth Greenhough, "Where Species Meet and Mingle: Endemic Human–Virus Relations, Embodied Communication and More-Than-Human Agency at the Common Cold Unit 1946–90," *Cultural Geographies* 19, no. 3 (2012): 281–301."

10. Alexis Zimmer, "The Disappearing Microbiota: the Coloniality of a Narrative and Anti-Colonial Proposals," *Environmental Humanities* 17, no. 2 (2025): 351–70.

11. For a review of the literature on this issue, see Jean-François Bach, "The Effect of Infections on Susceptibility to Autoimmune and Allergic Diseases," *New England Journal of Medicine* 347, no. 12 (2002): 911–20; and Jean-François Bach, "The Hygiene Hypothesis in Autoimmunity: The Role of Pathogens and Commensals," *Nature Reviews Immunology* 18, no. 2 (2018): 105–20.

12. See, for example, Gérard Eberl, "Immunity by Equilibrium," *Nature Reviews Immunology* 16, no. 8 (2016): 524–32. See also Lynn Chiu et al., "Protective Microbiota: From Localized to Long-Reaching Co-immunity," *Frontiers in Immunology* 8, no. 1678 (2017).

13. Bubbles of the most porous kind, as I will show in chapter 8.

14. Jean-Paul Pirnay et al., "The Phage Therapy Paradigm: Prêt-à-Porter or Sur-Mesure?," *Pharmaceutical Research* 28, no. 4 (2011): 934–37.

15. Shawna McCallin et al., "Metagenome Analysis of Russian and Georgian Pyophage Cocktails and a Placebo-Controlled Safety Trial of Single Phage Versus Phage Cocktail in Healthy *Staphylococcus Aureus* Carriers," *Environmental Microbiology* 20 (2018): 3278–93; Julia Villarroel et al., "Metagenomic Analysis of Therapeutic PYO Phage Cocktails from 1997 to 2014," *Viruses* 9, no. 11 (2017): 328; Henrike Zschach et al., "What Can We Learn from a Metagenomic Analysis of a Georgian Bacteriophage Cocktail?," *Viruses* 7, no. 12 (2015): 6570–89.

6. EVIDENCE-BASED MEDICINE

1. David Sackett et al., "Evidence-Based Medicine: What It Is and What It Isn't," *British Medical Journal* 312 (1996): 71–72.

2. Kirsten Bell, "Cochrane Reviews and the Behavioural Turn in Evidence-Based Medicine," *Health Sociology Review* 21, no. 3 (2012): 313–21;

Vincanne Adams, ed., *Metrics: What Counts in Global Health* (Duke University Press, 2016).

3. For a history of evidence-based medicine, see Harry Marks, *The Progress of Experiment: Science and Therapeutic Reform in the United States 1900–1990* (Cambridge University Press, 1997).

4. Monroe D. Eaton and Stanhope Bayne-Jones, "Bacteriophage Therapy," *Journal of the American Medical Association* 103 (1934): 1769–76; Albert Paul Krueger and Jane E. Scribner, "The Bacteriophage," *Journal of the American Medical Association* 116 (1941): 2160–69; Harry Morton and Frank Engley, "Dysentery Bacteriophage: Review of the Literature on Its Prophylactic and Therapeutic Uses in Man and in Experimental Infections in Animals," *Journal of the American Medical Association* 127, no. 10 (1945): 584–91.

5. Such data is worthy of historians' attention. In any case, they have not provided "evidence."

6. However, Alain Dublanchet told me that in the early years of the twenty-first century, a postoperative report on a patient with a severe foot infection described washing the wound before suturing it with a physiological liquid solution that contained phages obtained in Russia.

7. This book does not cover phage experimentation on animal models. I am currently conducting research on this subject while simultaneously studying the use of phages in animal health. This research will examine the involvement of animals in both situations, the questions asked of them, and the types of knowledge produced in these interactions.

8. See, for example, Marks, *The Progress of Experiment*; Stefan Timmermans and Marc Berg, *The Gold Standard: The Challenge of Evidence-Based Medicine and Standardization in Health Care* (Temple University Press, 2003); Philippe Pignarre, *Le Grand Secret de l'industrie pharmaceutique* (La Découverte, 2004). For a discussion of papers on various diseases, see Charlotte Brives et al., "What's in a Context? Tenses and Tensions in Evidence-Based Medicine," *Medical Anthropology: Cross-Cultural Studies in Health and Illness* 35, no. 5 (2016): 369–76.

9. For an examination of standardization and commensurability in clinical trials, see my work on HIV clinical trials: Charlotte Brives, "Identifying Ontologies in a Clinical Trial," *Social Studies of Science* 43, no. 3 (2013): 397–416; and Charlotte Brives, "The Myth of a Naturalised Male Circumcision: Heuristic Context and the Production of Scientific

Objects," *Global Public Health* 13, no. 11 (2018): 1599–1611. For a precise ethnographic and theoretical study of the transformation of pharmaceutical substances into pharmaceutical "objects" and what that implies, see Emilia Sanabria, *Plastic Bodies: Sex Hormones and Menstrual Suppression in Brazil* (Duke University Press, 2016).

10. Charlotte Brives, "Biomedical Packages: Adjusting Drugs, Bodies, and Environment in a Phase III Clinical Trial," *Medicine Anthropology Theory* 3, no. 1 (2016): 1–28.

11. A certain continuity with the structure of Priscilla Wald's "epidemic narratives" is apparent here, as well as with how these narratives systematically orient the responses to be provided: medicines, vaccines, barriers, and borders. This has led to measures such as favoring repeated vaccine doses and barrier measures to the detriment of massive reinvestment in the health care system. Narratives impose the question when, according to Stengers, the question should be created. Or, as I have written elsewhere, "living with" microbes is, in fact, simply a matter of "living in spite of them." See Charlotte Brives, "Pluribiose. Vivre avec les virus. Mais comment?," *Terrestres* 14 (2020).

12. Joseph Dumit and Emilia Sanabria, "Set, Setting and Clinical Trials: Colonial Technologies and Psychedelics," in *The Palgrave Handbook of the Anthropology of Technology*, ed. B. M. Hojer et al. (Palgrave Macmillan, 2022), 305.

13. Joseph Dumit, *Drugs for Life: How Pharmaceutical Companies Define Our Health* (Duke University Press, 2012).

14. See, for example, Marks, *The Progress of Experiment.*

15. Jeremy Greene tells the story behind the invention of three drugs for three conditions (Diuril for hypertension, Orinase for diabetes, and Mevacor for hypercholesterolemia) associated with specific disease risks. In the cases of hypertension and hypercholesterolemia, he describes how tests were used to keep lowering the thresholds at which individuals could be considered "at risk" and therefore "in need" of treatment. Each trial, each lowering of the threshold, corresponded to thousands of additional people being administered treatment. See Jeremy Greene, *Prescribing by Numbers: Drugs and the Definition of Disease* (Johns Hopkins University Press, 2007). See also Philippe Pignarre, *Comment la depression est devenue une épidémie* (La Découverte, 2012).

16. See Adriana Petryna, *When Experiments Travel: Clinical Trials and the Global Search for Human Subjects* (Princeton University Press, 2009), which describes, for example, how contract research organizations have gradually been entrusted with running clinical trials and the consequences of this subcontracting by the major pharmaceutical companies.

17. Concerning the ambivalence of clinical trials and their epistemologies, see Luc Berlivet and Ilana Löwy, "Hydroxychloroquine Controversies: Clinical Trials, Epistemology, and the Democratization of Science," *Medical Anthropology Quarterly* 34, no. 4 (2020): 525–41.

18. The PhagoBurn trial is registered in the EU Clinical Trials Register (EudraCT number 2014-000714-65) and in the US ClinicalTrials.gov registry (ID number NCT02116010).

19. Although clinical trials are generally conducted "double-blind," where neither the participant nor the prescriber knows the treatment actually taken to avoid biasing the results, the substantial difference in the presentation of the two treatments (phages in vials and the antibiotic cream in tubes) made it impossible to conceal the treatment allocation to the trial clinicians. The microbiologists responsible for processing the samples, however, were unaware of the randomization arm of the participants whose samples they were analyzing. For further details about this trial and to consult the available data, see Patrick Jault et al., "Efficacy and Tolerability of a Cocktail of Bacteriophages to Treat Burn Wounds Infected by *Pseudomonas aeruginosa* (PhagoBurn): A Randomised, Controlled, Double-Blind Phase 1/2 Trial," *Lancet Infectious Diseases* 19, no. 1 (2019): 35–45.

20. Jault et al. "Efficacy and Tolerability of a Cocktail of Bacteriophages," 35.

21. I will return to these aspects in subsequent chapters.

22. For detailed treatment reports, see Tristan Ferry et al., "Innovations for the Treatment of a Complex Bone and Joint Infection Due to XDR *Pseudomonas aeruginosa* including Local Application of a Selected Cocktail of Bacteriophages," *Journal of Antimicrobial Chemotherapy* 73, no. 10 (2018): 2901–903; Tristan Ferry et al., "Phage Therapy as Adjuvant to Conservative Surgery and Antibiotics to Salvage Patients with Relapsing *S. aureus* Prosthetic Knee Infection," *Frontiers in Medicine* 5, no. 7 (2020): 570–72; and Tristan Ferry et al., "Case Report: Arthroscopic 'Debridement Antibiotics and Implant Retention' with Local Injection

of Personalized Phage Therapy to Salvage a Relapsing *Pseudomonas aeruginosa* Prosthetic Knee Infection," *Frontiers in Medicine* 5, no. 8 (2021): 569159.

23. As this person wished to remain anonymous, I have used a pseudonym. Interview with Bérénice, June 2019.

24. European Medicines Agency, *Guideline on the Evaluation of Medicinal Products Indicated for Treatment of Bacterial Infections*, revision 3 (European Medicines Agency, 2019), 4.

25. Interview with Bérénice, June 2019.

26. Interview with Bérénice, June 2019.

7. ANTIBIOTIC INFRASTRUCTURES

1. Clare Chandler, "Current Accounts of Antimicrobial Resistance: Stabilisation, Individualisation and Antibiotics as Infrastructure." Palgrave Communications 5, no. 1 (2019):51 Susan Leigh Star, "The Ethnography of Infrastructure," *American Behavioral Scientist* 43, no. 3 (1999): 377–91. For an illustration of how infrastructures shape human lives without being solely material, see Geoffrey Bowker and Susan Leigh Star, *Sorting Things Out: Classification and Its Consequences* (MIT Press, 1999), in which the authors analyze the role of classificatory systems as infrastructures.

2. It is currently impossible to talk seriously about political ecology without considering the infrastructures inherited by humans. In this respect, this book shares the observations made by Alexandre Monnin, Diego Landivar, and Emmanuel Bonnet in *Héritage et Fermeture. Une écologie du démantèlement* (Divergences, 2021), in which the authors develop an "ecology of dismantling" and the notion of "ecological redirection."

3. Wai Chen, *Comment Fleming n'a pas inventé la pénicilline* (Les Empêcheurs de penser en rond, 1996), 57.

4. Fabrizio Spagnolo et al., "Why Do Antibiotics Exist?," *mBio* 12, no. 6 (2021): e01966–21. This is another example of the exclusion of a key female figure from the dominant narrative on penicillin.

5. For a more detailed history of penicillin, see Robert Bud, *Penicillin: Triumph and Tragedy* (Oxford University Press, 2007).

6. The military played a key role in the development of antibiotics, just as it is now playing a key role in the development of phage therapy. The PhagoBurn clinical trial was largely funded and supported by the

French Ministry of the Armed Forces. In Belgium, the team responsible for the development of phage therapy is based at the Queen Astrid Military Hospital. I will discuss this work in greater detail in chapter 9.

7. Scott H. Podolsky, "Antibiotics and the Social History of the Controlled Clinical Trial, 1950–1970," *Journal of the History of Medicine and Allied Sciences* 65, no. 3 (2010): 333.

8. For further details, see Scott Podolsky, *The Antibiotic Era: Reform, Resistance, and the Pursuit of a Rational Therapeutics* (Johns Hopkins University Press, 2015); and Harry Marks, *The Progress of Experiment: Science and Therapeutic Reform in the United States 1900–1990* (Cambridge University Press, 1997).

9. Claas Kirchhelle, *Pyrrhic Progress: The History of Antibiotics in Anglo-American Food Production* (Rutgers University Press, 2020).

10. Abigail Woods, "Decentring Antibiotics: UK Responses to the Diseases of Intensive Pig Production (ca. 1925–1965)," *Palgrave Communications* 5, no. 41 (2019).

11. Hannah Landecker, "The Food of Our Food: Medicated Feed and the Industrialization of Metabolism," in *Eating Beside Ourselves: Thresholds of Foods and Bodies*, ed. Heather Paxson (Duke University Press, 2023), 56–85.

12. Delphine Berdah, "Pour une autre histoire de la 'modernisation' des pratiques d'élevage. Des antibiotiques dans les rations pour stabiliser les systèmes sociotechniques pharmaceutiques français et britannique," in *Histoire des modernisations agricoles au xxᵉ siècle*, ed. M. Lyautey et al. (Presses universitaires de Rennes, 2021): 101–19. I am particularly indebted to Nicolas Fortané, a leading expert on antibiotic-related issues, who, upon reviewing the first draft of this chapter and the next, pointed out the ambiguities and controversies surrounding the types of knowledge produced at the time and how that knowledge was used by manufacturers to develop antibiotics.

13. Kirchhelle, *Pyrrhic Progress*. For more on the role of veterinarians, see Nicolas Fortané, "Veterinarian 'Responsibility': Conflicts of Definition and Appropriation Surrounding the Public Problem of Antimicrobial Resistance in France," *Palgrave Communications* 5, no. 67 (2019).

14. On the issue of livestock standardization, see, in particular, Ellen Silbergeld, *Chickenizing Farms and Food* (Johns Hopkins University Press, 2016); and Alex Blanchette, *Porkopolis: American Animality, Standardized Life, and the Factory Farm* (Duke University Press, 2020).

15. For a panorama of the studies conducted in that field, see the London Schol of Hygiene's "Antimicrobials in Society" website: https://www .lshtm.ac.uk/research/centres-projects-groups/amis-hub.

16. Laurie Denyer Willis and Clare Chandler, "Quick Fix for Care, Productivity, Hygiene and Inequality: Reframing the Entrenched Problem of Antibiotic Overuse," *BMJ Global Health* 4 (2019): e001590.

17. This type of argument is also developed in the seminal work of Susan Reynolds Whyte et al., *Social Lives of Medicine* (Cambridge University Press, 2002).

18. World Health Organization, *Global Action Plan on Antimicrobial Resistance* (World Health Organization, 2015).

19. Denyer Willis and Chandler, "Quick Fix for Care, Productivity, Hygiene and Inequality," 4.

20. Anna Tsing, *Proliférations* (Wildproject, 2022), 44.

21. Sidney W. Mintz, *Sweetness and Power: The Place of Sugar in Modern History* (Penguin, 1986).

22. Raj Patel and Jason W. Moore, *A History of the World in Seven Cheap Things: A Guide to Capitalism, Nature and the Future of the Planet* (University of California Press, 2018), 122.

23. Anna Tsing, *The Mushroom at the End of the World: On the Possibility of Life in Capitalist Ruins* (Princeton University Press, 2015), 78.

24. Blanchette, *Porkopolis*.

25. On shrimp farms in Bangladesh, see Steve Hinchliffe et al., "The AMR Problem: Demanding Economies, Biological Margins, and Co-producing Alternative Strategies," *Palgrave Communications* 4, no. 142 (2018). On citrus plantations in California, see Linda Nash, *Inescapable Ecologies: A History of Environment, Disease and Knowledge* (University of California Press, 2006).

26. For an overview of this issue, see, in particular, Christophe Bonneuil and Jean-Baptiste Fressoz, *The Shock of the Anthropocene* (Verso, 2016).

27. And the transition from biotic to abiotic. In this respect, it is sufficient to analogize that oil, like coal, is formed by the decomposition of residues of living organisms. Elizabeth Povinelli explores the distinction between what she calls "life" and "non-life," thereby proposing a critique of theories of biopolitical power, which, in her view, struggle to account for what is at stake in the most recent phases of the development of

liberalism. See Elizabeth A. Povinelli, *Geontologies: A Requiem to Late Liberalism* (Duke University Press, 2016).

28. Donna Haraway, "Anthropocene, Capitalocene, Plantationocene, Chthulucene: Making Kin," *Environmental Humanities* 6, no. 1 (2015): 159–65.

29. Andreas Malm was the first to use the term *Capitalocene*. However, unlike Moore, Malm tends to focus primarily on energy issues, which partly explains why his scope is different from those proposed by Moore and by proponents of *Plantationocene*. See Andreas Malm, *L'Anthropocène contre l'histoire* (La Fabrique, 2018); Jason Moore, "The Capitalocene, Part I: On the Nature and Origins of Our Ecological Crisis," *Journal of Peasant Studies* 44, no. 3 (2017): 594–630; Jason Moore, "The Capitalocene, Part II: Accumulation by Appropriation and the Centrality of Unpaid Work/ Energy," *Journal of Peasant Studies* 45, no. 2 (2018): 237–79; Jason Moore, *Capitalism in the Web of Life* (Verso, 2015).

30. Malcolm Ferdinand, on the other hand, uses the term *Negrocene* to refer to "the era in which the production of the Negro for the purpose of extending the colonial way of inhabiting played a fundamental role in the ecological and landscape changes of the Earth." See Malcom Ferdinand, *Une écologie décoloniale* (Le Seuil, 2019), 103.

31. According to Philippe Descola's terminology, another term for this is *naturalism*, which has been studied extensively by Bruno Latour. See Philippe Descola, *Beyond Nature and Culture* (University of Chicago Press, 2013); and Bruno Latour, *We Have Never Been Modern* (Harvard University Press, 1993).

32. Silvia Federici, *Caliban and the Witch: Women, the Body and Primitive Accumulation* (Penguin, 2017).

33. On women doing most of the work associated with the reproduction of the labor force, see Silvia Federici, *Le Capitalisme Patriarcal* (La Fabrique, 2019).

34. Patel and Moore, *A History of the World in Seven Cheap Things*, 31.

35. All the more so if we also see new territories as places to externalize the waste and pollution generated by these modes of capitalist relationships.

36. Will Steffen et al., "The Anthropocene: Are Humans Now Overwhelming the Great Forces of Nature?," *Ambio* 36, no. 8 (2007): 614–21;

John R. McNeill and Peter Engelke, *The Great Acceleration: An Environmental History of the Anthropocene Since 1945* (Harvard University Press, 2014).

37. In any case, I have never seen these chemical molecules included in the curves and diagrams used to show the characteristics of this period.

8. RECALCITRANCE AND FERALITY

The quote in the epigraph comes from Michelle Murphy, "Chemical Infrastructures of the Saint Clair River," in *Toxicants, Health and Regulation Since 1945*, ed. Soraya Boudia and Nathalie Jas (Pickering & Chatto, 2013), 105.

1. Alex Nading, *Mosquito Trails: Ecology, Health, and the Politics of Entanglement* (University of California Press, 2014).

2. Murphy, "Chemical Infrastructures," 115. Soraya Boudia and Nathalie Jas show how a mode of government of certain chemical substances based on mastery and control was quickly replaced by a series of accommodating arrangements to pursue the imperative of market construction. It was then a matter of proposing "a new social contract based on a partial recognition of risks and on new procedures of evaluation, management and reparation," See Soraya Boudia and Nathalie Jas, *Gouverner un monde toxique* (Quae, 2019), 58.

3. Anna Tsing, Jennifer Deger, Alder Keleman Saxena, and Feifei Zhou, along with dozens of other researchers, have made this the subject of the web-based project "Feral Atlas"; see https://feralatlas.org/.

4. The literature first mentioned resistance to salvarsan, a product used to treat syphilis between 1910 and 1940, in 1924, according to Dov Stekel, "First Report of Antimicrobial Resistance Pre-dates Penicillin," *Nature* 562, no. 7726 (2018): 192.

5. Alexander Fleming, "Penicillin," Nobel Lecture, December 11, 1945.

6. For more on the development of antibiotic resistance as a problem, see Christoph Gradmann, "From Lighthouse to Hothouse: Hospital Hygiene, Antibiotics and the Evolution of Infectious Disease, 1950–1990," *History and Philosophy of Life Sciences* 40, no. 1 (2018).

7. Angela Creager, *The Life of a Virus: Tobacco Mosaic Virus as an Experimental Model, 1930–1965* (University of Chicago Press, 2002).

8. Ephraim S. Anderson, "Drug Resistance in *Salmonella typhimurium* and Its Implications," *British Medical Journal* 3 (1968), 333–39.
9. Scott Podolsky et al., "History Teaches Us That Confronting Antibiotic Resistance Requires Stronger Global Collective Action," *Journal of Law, Medicine and Ethics* 43, no. 2 (2015): 27–32.
10. Scott Podolsky, "The Evolving Response to Antibiotic Resistance (1945–2018)," *Palgrave Communications* 4, no. 124 (2018).
11. In this book, I do not dwell on the "One Health" concept, which is sometimes used by phage therapy actors seeking to emphasize the ecological dimension of the infectious problems they face. This concept poses a number of problems in its fundamental principles and the types of actions it seeks to promote. For further details, see, for example, Steve Hinchliffe, "More Than One World, More Than One Health: Re-configuring Interspecies Health," *Social Science & Medicine* 129 (2015): 28–35; and Nicolas Lainé and Serge Morand, "Linking Humans, Their Animals, and the Environment *Again*: A Decolonized and More-Than-Human Approach to 'One Health,'" *Parasite* 27, no. 55 (2020).
12. Sally C. Davies, *Annual Report of the Chief Medical Officer*, vol. 2, 2011, *Infections and the Rise of Antimicrobial Resistance* (Department of Health, 2013), 16.
13. Hannah Landecker, "Antibiotic Resistance and the Biology of History," *Body & Society* 22, no. 4 (2016): 19–52.
14. Eugene Koonin et al., "Comparison of Phylogenetic Trees and Search for a Central Trend in the 'Forest of Life,'" *Journal of Computational Biology* 18, no. 7 (2011): 917–24.
15. Once again, tribute should be paid to the pioneering work of Lynn Margulis. See Lynn Sagan, "On the Origin of Mitosing Cells," *Journal of Theoretical Biology* 14, no. 3 (1967): 225–74; and Lynn Margulis and Dorion Sagan, *Microcosmos: Four Billion Years of Microbial Evolution* (University of California Press, 1997). For a comprehensive review of the various theories in microbiology, see Maureen O'Malley, *Philosophy of Microbiology* (Cambridge University Press, 2014).
16. On the question of the production and circulation of ignorance in public health, including how the very structuring of knowledge invisibilizes certain relationships, see Emilia Sanabria, "Circulating Ignorance: Complexity and Agnogenesis in the Obesity 'Epidemics,'" *Cultural

Anthropology 31, no. 1 (2016): 131–58. On agnotology, see Robert N. Proctor and Londa Schiebinger, eds., *Agnotology: The Making and Unmaking of Ignorance* (Stanford University Press, 2008).

17. However, an ever-increasing number of opinion pieces and papers are being published on the actions that must be taken. See, for example, Podolsky et al., "History Teaches Us That Confronting Antibiotic Resistance Requires Stronger Global Collective Action."

18. Podolsky, "The Evolving Response to Antibiotic Resistance."

19. I would like to thank Rémy Froissart for summarizing and presenting the elements of this box in such an accessible manner, clearly explaining what is covered by the term *microbial evolution*.

20. Alison H. Holmes et al., "Understanding the Mechanisms and Drivers of Antimicrobial Resistance," *Lancet* 387, no. 10014 (2016): 176–87; Julian Davies and Dorothy Davies, "Origins and Evolution of Antibiotic Resistance," *Microbiology and Molecular Biology Review* 74, no. 3 (2010): 417–33; Liselotte Diaz Högberg et al., "The Global Need for Effective Antibiotics: Challenges and Recent Advances," *Trends in Pharmacological Sciences* 31, no. 11 (2010): 509–15.

21. Gérard Moulin et al., "A Comparison of Antimicrobial Usage in Human and Veterinary Medicine in France from 1999 to 2005," *Journal of Antimicrobial Chemotherapy* 62, no. 3 (2008): 617–25; Karen L. Tang et al., "Restricting the Use of Antibiotics in Food-Producing Animals and Its Associations with Antibiotic Resistance in Food-Producing Animals and Human Beings: A Systematic Review and Meta-analysis," *Lancet Planetary Health* 1, no. 8 (2017): e316–27.

22. Alison H. Holmes et al., "Understanding the Mechanisms and Drivers of Antimicrobial Resistance," *Lancet* 387, no. 10014 (2016): 176–87.

23. Michael Gillings, "Evolutionary Consequences of Antibiotic Use for the Resistome, Mobilome, and Microbial Pangenome," *Frontiers in Microbiology* 4, no. 4 (2013).

24. Landecker, "Antibiotic Resistance and the Biology of History"; Pierre-Edouard Fournier et al., "Comparative Genomics of Multidrug Resistance in *Acinetobacter baumannii*," *PLoS Genetics* 2, no. 1 (2006): e7.

25. The infectiology of injuries caused by explosions, sometimes accompanied by "human shrapnel"—bone fragments from destroyed bodies penetrating other bodies—differs greatly from that of gunshot wounds,

for example. For more on the role of heavy metals in the acquisition of resistance in Iraqibacter, see Wael Bazzi et al., "Heavy Metal Toxicity in Armed Conflicts Potentiates AMR in *A. baumannii* by Selecting for Antibiotic and Heavy Metal Co-resistance Mechanisms," *Frontiers in Microbiology* 11, no. 68 (2020).

26. Omar Dewachi, "Iraqibacter and the Pathologies of Intervention," *Middle East Report* 290 (2019).

27. Landecker, "Antibiotic Resistance and the Biology of History," 3.

28. Although I have been unable to verify this assertion, John McNeill and his colleagues would appear to have drawn on Karl Polanyi's work in coining the term "great acceleration."

29. Murphy, "Chemical Infrastructures."

30. Alexis Zimmer, *Brouillards toxiques. Vallée de la Meuse, 1930, contre-enquête* (Zones Sensibles, 2016), 221–22.

31. See, in particular, Max Liboiron, *Pollution Is Colonialism* (Duke University Press, 2021).

9. TOWARD A PLURIBIOTIC MODEL?

1. Thom van Dooren et al., "Multispecies Studies: Cultivating Arts of Attentiveness," *Environmental Humanities* 8, no. 1 (2016): 1–23.

2. Émilie Hache, *Ce à quoi nous tenons. Propositions pour une écologie pragmatique* (La Découverte, 2011), 98–99.

3. Hache, *Ce à quoi nous tenons*, 13.

4. Maya Merabishvili et al., "Quality-Controlled Small-Scale Production of a Well-Defined Bacteriophage Cocktail for Use in Human Clinical Trials," *PLoS One* 4, no. 3 (2009): e4944.

5. This is why most of the therapeutic treatments involving bacteriophage viruses are carried out within the framework of Article 37 of the Declaration of Helsinki.

6. Gilbert Verbeken et al., "European Regulatory Conundrum of Phage Therapy," *Future Microbiology* 2, no. 5 (2007): 485–91.

7. Alan Fauconnier, "Phage Therapy Regulation: From Night to Dawn," *Viruses* 11, no. 4 (2019): 352.

8. Laurent Debarbieux et al., "A Bacteriophage Journey at the European Medicines Agency," *FEMS Microbiology Letters* 363, no. 2 (2016): fnv225.

9. Opportunities for adaptation do exist, including, as discussed in chapter 6, the opportunity for trials to evaluate not a given phage but a therapeutic regimen using phages in the context of a specific medical condition.

10. Interview with Gilbert Verbeken, February 19, 2018.

11. This is also the perspective from which I wrote this book.

12. Vincanne Adams, ed., *Metrics: What Counts in Global Health* (Duke University Press, 2016), 38.

13. As I mentioned in chapter 6, most offshored trials are carried out by contract research organizations. Harmonization of regulations mainly occurred in Latin American and Eastern European countries in the 1990s. India's alignment with these regulations occurred only in 2005, when it became a key location for clinical trials for major pharmaceutical groups. See Kaushik Sunder Rajan, *Pharmocracy: Value, Politics, and Knowledge in Global Medicine* (Duke University Press, 2017).

14. For example, see Jeremy Greene et al., eds., *Therapeutic Revolutions: Pharmaceuticals and Social Change in the Twentieth Century* (University of Chicago Press, 2016). Work on the *Diagnostic and Statistical Manual of Mental Disorders* should also be mentioned here, such as that described in Ian Hacking, *Mad Travelers: Reflections on the Reality of Transient Mental Illness* (University of Virginia Press, 1998).

15. For a detailed discussion of patents and their role in price formulation, see Maurice Cassier, "Patents and Public Health in France: Pharmaceutical Patent Law In-the-Making at the Patent Office Between the Two World Wars," *History and Technology* 24, no. 2 (2008): 135–51; and Maurice Cassier, "Value Regimes and Pricing in the Pharmaceutical Industry: Financial Capital Inflation (Hepatitis C) Versus Innovation and Production Capital Savings for Malaria Medicines," *BioSocieties* 16, no. 3 (2021): 323–41.

16. Antibiotics have played an important role in both factors, as discussed in chapter 5.

17. See figure 8.1, which shows the emergence of resistance to various antibiotics.

18. Pherecydes Pharma and other start-ups have actually filed a number of patents, but these have concerned not the patenting of a given phage but rather combinations of various phages. Further analysis will

be required to understand both what is patented and the role of these patents in companies' development models. Appropriation strategies can involve the patenting of production or purification processes, for example. The stakes are high, and everything is still in progress, which is why I will not discuss these issues further here.

19. For a detailed study of the commodification of blood and stem cell banks, see Catherine Waldby and Robert Mitchell, *Tissue Economies: Blood, Organs, and Cell Lines in Late Capitalism* (Duke University Press, 2006).

20. These two intertwined issues will be the subject of further research. For a legal approach, see Florence Bellivier and Christine Noiville, eds., *Contrats et Vivant. Le droit de la circulation des ressources biologiques* (LGDJ, 2006); and Marie-Angèle Hermitte, *L'Emprise des droits intellectuels sur le monde vivant* (Quae, 2016). The patentability of living organisms has been the subject of numerous studies, including Fabienne Orsi, "La constitution d'un nouveau droit de propriété intellectuelle sur le vivant aux États-Unis: origine et signification économique d'un dépassement de frontière," *Revue d'économie industrielle* 99 (2002): 65–86. For a historical analysis of the links between the production of knowledge and the patenting of living organisms, see Christophe Bonneuil and Frédéric Thomas, *Gènes, pouvoirs et profits. Recherche publique et régimes de production des savoirs de Mendel aux OGM* (Quae, 2009).

21. Nevertheless, some of the people I encountered are not opposed to private investment, provided that it is controlled and enables the development of sur-mesure phage therapy.

22. *Directive 2001/83/EC of the European Parliament and of the Council of 6 November 2001 on the Community Code Relating to Medicinal Products for Human Use,* Article 3.

23. The reality is even more complex, however, as the very content of the monograph, that is, the production standards, are themselves the subject of negotiations between the various actors and regulatory agencies. Once again, the latter do not produce regulations, or even norms and standards. They simply apply them or validate the propositions made by the various stakeholders on the basis of existing texts.

24. This is a lengthy regulatory undertaking, as described in Jean-Paul Pirnay et al., "The Magistral Phage," *Viruses* 10, no. 2 (2018).

25. For a detailed analysis of these requests and their handling, see Sarah Djebara et al., "Processing Phage Therapy Requests in a Brussels Military Hospital: Lessons Identified," *Viruses* 11 (2019): 265.

26. Since the publication of the French edition, phage therapy in Belgium has undergone a number of new developments. The QAMH team partnered with the local private sector, which initially allowed for the outsourcing of logistics, customs management, and regulatory aspects. But the private company eventually created its own spin-off, insisting on GMP-certification with the aim of commercialization, breaking away from the project developed by the QAMH team, which explicitly rejected the marketing authorization model. The Belgian team at QAMH, for its part, continues with its non GMP-certified phage-based strategy, compatible with the personalized phage therapy approach. They have published a landmark publication in *Nature*, in which they review the first one hundred patients treated with phage APIs provided by QAMH since the start of the venture. J.-P. Pirnay, S. Djebara, G. Steurs, J. Griselain, C. Cochez, S. De Soir, S., et al. "Personalized bacteriophage therapy outcomes for 100 consecutive cases: A multicentre, multinational, retrospective observational study," *Nature Microbiology* 9, no. 6 (2024): 1434–53, https://doi.org/10.1038/s41564-024 -01705-x. For more details, see Kameda & Brives (forthcoming).

27. The list of experts appointed and speakers heard, as well as the minutes of the 2016 and 2019 CSSTs, can be consulted on the ANSM website: https://ansm.sante.fr/actualites/phagotherapie-publication-du-compte -rendu-du-csst-phagotherapie-retour-dexperience-et-perspectives.

28. Agence Nationale de Sécurité du Médicament (ANSM), *Compte rendu de séance: Comité scientifique spécialisé temporaire. Phagothérapie – retour d'expérience et perspectives* (ANSM, 2019), 1.

29. ANSM, *Compte rendu de séance*, 10.

30. The start-up made this decision based on the stability problems encountered with phage cocktails in the PhagoBurn trial (see chapter 3).

31. Since then, the start-up has been bought out but finally filed for bankruptcy in the spring of 2025. For more details about the recent developments of phage therapy, see Kameda & Brives (forthcoming).

32. ANSM, *Compte rendu de séance*, 10.

33. ANSM, *Compte rendu de séance*, 15.

34. ANSM, *Compte rendu de séance*, 16.

35. The day before, the Arte television channel had broadcast a documentary about antibiotic resistance and had invited Professor Didier Raoult to take part in an ensuing debate in prime time. With characteristic aplomb, he presented highly debatable arguments to explain that antibiotic resistance was a nonissue. Raoult's comments on antibiotics and his position on phage therapy fueled discussion during meeting breaks.

36. By May 2022, the Lyon-based team had treated thirty-eight people with phages, mainly for osteoarticular infections but also for endocarditis and pneumonia. The number of requests for treatment and eligible candidates is constantly increasing.

37. With the exception of the Agence Générale des Equipements et Produits de Santé, the in-house pharmacy of the Assistance publique—Hôpitaux de Paris, which was one of the last pharmacies to benefit from this status, now reserved exclusively for private companies.

38. They could also be used in clinical trials if the trials were carried out in the same establishment.

39. Article L. 5121–1, paragraph 2, of the French Public Health Code stipulates that the qualification of "hospital preparation" applies only as long as no proprietary pharmaceutical product is available.

40. In this regard, the story of Zolgensma, a product designed to treat children with spinal muscular atrophy (a rare disease leading to early death), is edifying. The drug was first developed in public laboratories, partly financed by Genethon (and therefore by telethon donations), before being taken over by the start-up AveXis, which was subsequently bought out by Novartis. Today, the cost of one treatment of Zolgensma is almost €2 million.

41. On the mobilization of patients in the fight against HIV, see, for example, Steven Epstein, *Impure Science: AIDS, Activism, and the Politics of Knowledge* (University of California Press, 1996).

42. Cassier, "Value Regimes and Pricing in the Pharmaceutical Industry," 1.

43. For more on the Brazilian initiative, see Maurice Cassier and Marilena Correa, "Éloge de la copie: le *reverse-engineering* des antirétroviraux contre le VIH/sida dans les laboratoires pharmaceutiques brésiliens," *Sciences sociales et santé* 27, no. 3 (2009): 77–103.

44. This initiative, created by Médecins Sans Frontières and the Pasteur Institute in response to the withdrawal of major R&D laboratories from the tropical disease field, brings together a wide variety of partners, from the Oswaldo-Cruz Foundation in Brazil, the Kenyan Medical Research Institute, and the Indian Council for Medical Research to the World Health Organization, the United Nations Development Program, and the World Bank.

45. Cassier, "Value Regimes and Pricing in the Pharmaceutical Industry." However, this type of arrangement remains an exception to the rule, enabling the participating companies to improve their image at little cost. For more on the production of ASAQ in Sub-Saharan Africa, see, for example, Jessica Pourraz, "Produire des médicaments en Afrique subsaharienne à l'heure de la santé globale. Le cas des antipaludiques au Ghana," *Politique africaine* 156, no. 4 (2019): 41–60.

46. However, it should be remembered that alternative antibiotic development models have been proliferating in recent years in an attempt to address the problems posed by antibiotic resistance.

47. Although I have not had an opportunity to return to Switzerland to follow the latest developments in Grégory Resch's work, the same type of initiative seems to be occurring there, at the Lausanne University Hospital.

BIBLIOGRAPHY

Adams, Vincanne, ed. *Metrics: What Counts in Global Health*. Duke University Press, 2016.

Agence nationale de sécurité du médicament (ANSM). *Compte rendu de séance: Comité scientifique spécialisé temporaire. Phagothérapie—retour d'expérience et perspectives*. ANSM, 2019.

Anderson, Ephraim S. "Drug Resistance in *Salmonella typhimurium* and Its Implications." *British Medical Journal* 3, no. 5614 (1968): 333–39.

Anderson, Warwick. "Natural Histories of Infectious Disease: Ecological Vision in Twentieth-Century Biomedical Science." *Osiris* 19 (2004): 39–61.

Ankeny, Rachel. "The Conqueror Worm: An Historical and Philosophical Examination of the Use of the Nematode *Caenorhabditis elegans* as a Model Organism." PhD diss., University of Pittsburgh, 1997.

Ansaldi, Mireille, Pascale Boulanger, Charlotte Brives, et al. "Les applications antibactériennes des bacteriophages." *Virologie* 24, no. 1 (2020): 23–36.

——. "Un siècle de recherche sur les bacteriophages." *Virologie* 24, no. 1 (2020): 9–22.

Antimicrobial Resistance Collaborators. "Global Burden of Bacterial Antimicrobial Resistance in 2019: A Systematic Analysis." *Lancet* 399, no. 10325 (2022): 629–55.

Bach, Jean-François. "The Effect of Infections on Susceptibility to Autoimmune and Allergic Diseases." *New England Journal of Medicine* 347, no. 12 (2002): 911–20.

——. "The Hygiene Hypothesis in Autoimmunity: The Role of Pathogens and Commensals." *Nature Reviews Immunology* 18, no. 2 (2018): 105–20.

Bapteste, Éric. *Tous entrelacés! Des gènes aux super-organismes: les réseaux de l'évolution.* Belin, 2017.

Barad, Karen. *Meeting the Universe Halfway: Quantum Physics and the Entanglement of Matter and Meaning.* Duke University Press, 2007.

Bazzi, Wael, Antoine G. Abou Fayad, Aya Nasser, et al. "Heavy Metal Toxicity in Armed Conflicts Potentiates AMR in *A. baumannii* by Selecting for Antibiotic and Heavy Metal Co-resistance Mechanisms." *Frontiers in Microbiology* 11, no. 68 (2020).

Bell, Kirsten. "Cochrane Reviews and the Behavioural Turn in Evidence-Based Medicine." *Health Sociology Review* 21, no. 3 (2012): 313–21.

Bellivier, Florence, and Christine Noiville, eds. *Contrats et Vivant. Le droit de la circulation des ressources biologiques.* LGDJ, 2006.

Benezra, Amber. "Race in the Microbiome." *Science, Technology & Human Values* 45, no. 5 (2020): 877–902.

Benezra, Amber, Joseph DeStefano, and Jeffrey Gordon. "Anthropology of Microbes." *Proceedings of the National Academy of Sciences* 109, no. 17 (2012): 6378–81.

Bensaude-Vincent, Bernadette. "Guerre et paix avec le coronavirus." *Terrestres* 13 (2020).

Berdah, Delphine. "Pour une autre histoire de la 'modernisation' des pratiques d'élevage. Des antibiotiques dans les rations pour stabiliser les systèmes sociotechniques pharmaceutiques français et britannique," in *Histoire des modernisations agricoles au xxe siècle*, ed. M. Lyautey, L. Humbert, and C. Bonneuil. Presses universitaires de Rennes, 2021.

Berlivet, Luc, and Ilana Löwy. "Hydroxychloroquine Controversies: Clinical Trials, Epistemology, and the Democratization of Science." *Medical Anthropology Quarterly* 34, no. 4 (2020): 525–41.

Betts, Alex, Oliver Katz, and Michael E. Hochberg. "Contrasted Coevolutionary Dynamics Between a Bacterial Pathogen and Its Bacteriophages." *Proceedings of the National Academy of Sciences* 111, no. 30 (2014).

Blanchette, Alex. *Porkopolis: American Animality, Standardized Life, and the Factory Farm.* Duke University Press, 2020.

Blaser, Martin. *Missing Microbes: How the Overuse of Antibiotics Is Fueling Our Modern Plagues.* Picador, 2014.

Bonneuil, Christophe, and Jean-Baptiste Fressoz. *The Shock of the Anthropocene.* Verso, 2016.

Bonneuil, Christophe, and Frédéric Thomas. *Gènes, pouvoirs et profits. Recherche publique et régimes de production des savoirs de Mendel aux OGM.* Quae, 2009.

Boudia, Soraya, and Nathalie Jas. *Gouverner un monde toxique.* Quae, 2019.

Bourgeois, Gaëlle. "Histoire de la phagothérapie à Lyon." MD thesis, Université Claude-Bernard Lyon 1, 2020.

Bowker, Geoffrey, and Susan Leigh Star. *Sorting Things Out: Classification and Its Consequences.* MIT Press, 1999.

Brives, Charlotte. "Biomedical Packages: Adjusting Drugs, Bodies, and Environment in a Phase III Clinical Trial." *Medicine Anthropology Theory* 3, no. 1 (2016): 1–28.

——. "From Fighting Against to Becoming With: Viruses as Companion Species." In *Hybrid Communities: Biosocial Approaches to Domestication and Other Trans-species Relationships*, ed C. Stépanoff and J.-D. Vigne. Routledge, 2017.

——. "Identifying Ontologies in a Clinical Trial." *Social Studies of Science* 43, no. 3 (2013): 397–416.

——. "Des levures et des hommes. Anthropologie des relations entre humains et non humains dans un laboratoire de biologie." PhD diss., Université Victor-Segalen, 2010.

——. "The Myth of a Naturalised Male Circumcision: Heuristic Context and the Production of Scientific Objects." *Global Public Health* 13, no. 11 (2018): 1599–1611.

——. "Pluribiose. Vivre avec les virus. Mais comment?" *Terrestres* 14 (2020).

——. "Pluribiosis and the Never-Ending Microgeohistories." In *With Microbes*, ed. Charlotte Brives, Matthäus Rest, and Salla Sariola. Mattering, 2021.

——. "The Politics of Amphibiosis: The War Against Viruses Will Not Take Place." *Somatosphere: Science, Medicine, and Anthropology*, April 19, 2020.

——. "Le rôle des écrits éphémères dans la production des faits scientifiques: la domestication de la levure sauvage." *Langage et société* 1, no. 127 (2009): 71–81.

Brives, Charlotte, Frédéric Le Marcis, and Emilia Sanabria. "What's in a Context? Tenses and Tensions in Evidence-Based Medicine." *Medical Anthropology: Cross-Cultural Studies in Health and Illness* 35, no. 5 (2016): 369–76.

Brives, Charlotte, Matthäus Rest, and Salla Sariola, eds. *With Microbes*. Mattering, 2021.

Brives, Charlotte, and Alexis Zimmer. "Ecologies and Promises of the Microbial Turn." Special issue, *Revue d'anthropologie des connaissances* 15, no. 3 (2021).

Brüssow, Harald, Carlos Canchaya, and Wolf-Dietrich Hardt. "Phages and the Evolution of Bacterial Pathogens: From Genomic Rearrangements to Lysogenic Conversion." *Microbiology and Molecular Biology Reviews* 68, no. 3 (2004): 560–602.

Bud, Robert. *Penicillin: Triumph and Tragedy*. Oxford University Press, 2007.

Cassier, Maurice. "Patents and Public Health in France: Pharmaceutical Patent Law In-the-Making at the Patent Office Between the Two World Wars." *History and Technology* 24, no. 2 (2008): 135–51.

——. "Value Regimes and Pricing in the Pharmaceutical Industry: Financial Capital Inflation (Hepatitis C) Versus Innovation and Production Capital Savings for Malaria Medicines." *BioSocieties* 16, no. 3 (2021): 323–41.

Cassier, Maurice, and Marilena Correa. "Éloge de la copie: le reverse-engineering des antirétroviraux contre le VIH/sida dans les laboratoires pharmaceutiques brésiliens." *Sciences sociales et santé* 27, no. 3 (2009): 77–103.

Chan, Benjamin, Mark Sistrom, John Wertz, et al. "Phage Selection Restores Antibiotic Sensitivity in MDR *Pseudomonas aeruginosa*." *Scientific Reports* 6, no. 26717 (2016).

Chandler, Clare. "Current Accounts of Antimicrobial Resistance: Stabilisation, Individualisation and Antibiotics as Infrastructure." *Palgrave Communications* 5, no. 1 (2019): 53.

Chartier, Denis. "The Deplantationocene: Listening to Yeasts and Rejecting the Plantation Worldview." In *With Microbes*, ed. Charlotte Brives, Matthäus Rest, and Salla Sariola. Mattering, 2021.

Chen, Wai. *Comment Fleming n'a pas inventé la pénicilline*. Les Empêcheurs de penser en rond, 1996.

Chiu, Lynn, Thomas Bazin, Marie-Elise Truchetet, Thierry Schaeverbeke, Laurence Delhaes, and Thomas Pradeu. "Protective Microbiota: From Localized to Long-Reaching Co-immunity." *Frontiers in Immunology* 8, no. 1678 (2017).

Clarke, Adele, and Joan Fujimura, eds. *The Right Tools for the Job: At Work in Twentieth-Century Life Sciences*. Princeton University Press, 1992.

Clause, Bonnie. "The Wistar Rat as a Right Choice: Establishing Mammalian Standards and the Ideal of a Standardized Mammal." *Journal of the History of Biology* 26, no. 2 (1993): 329–49.

Cooper, Melinda. *Life as Surplus: Biotechnology and Capitalism in the Neoliberal Era*. University of Washington Press, 2008.

Creager, Angela. *The Life of a Virus: Tobacco Mosaic Virus as an Experimental Model, 1930–1965*. University of Chicago Press, 2002.

Czaplewski, Lloyd, Richard Bax, Martha Clokie, et al. "Alternatives to Antibiotics—a Pipeline Portfolio Review." *Lancet Infectious Diseases* 16, no. 2 (2016): 239–51.

Davies, Julian, and Dorothy Davies. "Origins and Evolution of Antibiotic Resistance." *Microbiology and Molecular Biology Review* 74, no. 3 (2010): 417–33.

Davies, Sally C. *Annual Report of the Chief Medical Officer*. Vol. 2, 2011, *Infections and the Rise of Antimicrobial Resistance*. Department of Health, 2013.

De Freitas Almeida, Gabriel Magno, and Lotta-Riina Sundberg. "The Forgotten Tale of Brazilian Phage Therapy." *Lancet Infectious Diseases* 20, no. 5 (2020): e90–e101.

Debarbieux, Laurent, Jean-Paul Pirnay, Golbert Verbeken, et al. "A Bacteriophage Journey at the European Medicines Agency." *FEMS Microbiology Letters* 363, no. 2 (2016): fnv225.

Denyer Willis, Laurie, and Clare Chandler. "Quick Fix for Care, Productivity, Hygiene and Inequality: Reframing the Entrenched Problem of Antibiotic Overuse." *BMJ Global Health* 4 (2019): e001590.

Descola, Philippe. *Beyond Nature and Culture*. University of Chicago Press, 2013.

Despret, Vinciane. *La danse du cratérope écaillé. Naissance d'une théorie éthologique*. Les Empêcheurs de penser en rond, 1996.

Despret, Vinciane. "En finir avec l'innocence, dialogue avec Isabelle Stengers et Donna Haraway." In *Penser avec Donna Haraway*, ed. E. Dorlin and E. Rodriguez. Puf, 2012.

Dewachi, Omar. "Iraqibacter and the Pathologies of Intervention." *Middle East Report* 290 (2019).

D'Hérelle, Félix. "Sur un microbe invisible antagoniste des bacilles dysentériques." *Comptes rendus de l'Académie des sciences* 165, no. 11 (1917): 373–75.

Diaz Högberg, Liselotte, Andreas Heddini, and Otto Cars. "The Global Need for Effective Antibiotics: Challenges and Recent Advances." *Trends in Pharmacological Sciences* 31, no. 11 (2010): 509–15.

Djebara, Sarah, Christiane Maussen, Daniel De Vos, et al. "Processing Phage Therapy Requests in a Brussels Military Hospital: Lessons Identified." *Viruses* 11 (2019): 265.

Dublanchet, Alain. *Autobiographie de Félix d'Hérelle. Les pérégrinations d'un bactériologiste.* Éditions médicales internationals, 2017.

——. *La Phagothérapie. Des virus pour combattre les infections.* Favre, 2017.

Dufour, Nicolas, Raphaëlle Delattre, Anne Chevallereau, Jean-Damien Ricard, and Laurent Debarbieux. "Phage Therapy of Pneumonia Is Not Associated with an Overstimulation of the Inflammatory Response Compared to Antibiotic Treatment in Mice." *Antimicrobial Agents and Chemotherapy* 63, no. 8 (2019): e00379–19.

Dufour, Nicolas, Raphaëlle Delattre, Jean-Damien Ricard, and Laurent Debarbieux. "The Lysis of Pathogenic *Escherichia coli* by Bacteriophages Releases Less Endotoxin Than by β-Lactams." *Clinical Infectious Diseases* 64, no. 11 (2017): 1582–88.

Dumit, Joseph. *Drugs for Life: How Pharmaceutical Companies Define Our Health.* Duke University Press, 2012.

Dumit, Joseph, and Emilia Sanabria. "Set, Setting and Clinical Trials: Colonial Technologies and Psychedelics." In *The Palgrave Handbook of the Anthropology of Technology*, ed B. M. Hojer, A. Wahlberg, R. Douglas-Jones, C. Hasse, K. Hoeyer, D. Brogård Kristensen, and B. R. Winthereik. Palgrave Macmillan, 2022.

Dupré, John. *Processes of Life: Essays in the Philosophy of Biology.* Oxford University Press, 2011.

Dupré, John, and Stephan Guttinger. "Viruses as Living Processes." *Studies in History and Philosophy of Biological and Biomedical Sciences* 59 (2016): 109–16.

Dupré, John, and Maureen O'Malley. "Varieties of Living Things: Life at the Intersection of Lineage and Metabolism." *Philosophy & Theory in Biology* 1, no. 3 (2009).

Dupressoir, Anne, and Thierry Heidmann. "Les syncytines, des protéines d'enveloppe rétrovirales capturées au profit du développement placentaire." *Médecine/Sciences* 27, no. 2 (2011): 163–69.

Eaton, Monroe D., and Stanhope Bayne-Jones. "Bacteriophage Therapy." *Journal of the American Medical Association* 103 (1934): 1769–76.

Eberl, Gérard. "Immunity by Equilibrium." *Nature Reviews Immunology* 16, no. 8 (2016): 524–32.

Epstein, Steven. *Impure Science: AIDS, Activism, and the Politics of Knowledge.* University of California Press, 1996.

European Medicines Agency. *Guideline on the Evaluation of Medicinal Products Indicated for Treatment of Bacterial Infections*, revision 3. European Medicines Agency, 2019.

Fauconnier, Alan. "Phage Therapy Regulation: From Night to Dawn." *Viruses* 11, no. 4 (2019): 352.

Federici, Silvia. *Caliban and the Witch: Women, the Body and Primitive Accumulation.* Penguin, 2017.

——. *Le Capitalisme patriarcal.* La Fabrique, 2019.

Ferdinand, Malcom. *Une écologie décoloniale.* Le Seuil, 2019.

Ferry, Tristan, Fabien Boucher, Cindy Fèvre, et al. "Innovations for the Treatment of a Complex Bone and Joint Infection Due to XDR *Pseudomonas aeruginosa* Including Local Application of a Selected Cocktail of Bacteriophages." *Journal of Antimicrobial Chemotherapy* 73, no. 10 (2018): 2901–903.

Ferry, Tristan, Camille Kolenda, Cécile Batailler, et al. "Case Report: Arthroscopic 'Debridement Antibiotics and Implant Retention' with Local Injection of Personalized Phage Therapy to Salvage a Relapsing *Pseudomonas aeruginosa* Prosthetic Knee Infection." *Frontiers in Medicine* 5, no. 8 (2021): 569159.

Ferry, Tristan, Camille Kolenda, Cécile Batailler, et al. "Phage Therapy as Adjuvant to Conservative Surgery and Antibiotics to Salvage Patients with Relapsing *S. aureus* Prosthetic Knee Infection." *Frontiers in Medicine.* 5, no. 7 (2020): 570–72.

Fleming, Alexander. "Penicillin." Nobel Lecture, December 11, 1945.

Fortané, Nicolas. "Veterinarian 'Responsibility': Conflicts of Definition and Appropriation Surrounding the Public Problem of Antimicrobial Resistance in France." *Palgrave Communications* 5, no. 67 (2019).

Fortun, Kim. "Ethnography in Late Industrialism." *Cultural Anthropology* 27, no. 3 (2012): 446–64.

Fournier, Pierre-Edouard, David Vallenet, Valérie Barbe, et al. "Comparative Genomics of Multidrug Resistance in *Acinetobacter baumannii*." *PLoS Genetics* 2, no. 1 (2006): e7.

Fox Keller, Evelyn. *The Century of the Gene.* Harvard University Press, 2002.

——. *Making Sense of Life: Explaining Biological Development with Models, Metaphors, and Machines*. Harvard University Press, 2002.

Fruciano, Emiliano. "La Phagothérapie. Émergence d'une idée controversée et logique d'un échec (1917–1949)." PhD diss., École des hautes études en sciences sociales, 2011.

Furfaro, Lucy, Matthew Payne, and Barbara Chang. "Bacteriophage Therapy: Clinical Trials and Regulatory Hurdles." *Frontiers in Cellular and Infection Microbiology* 8, no. 376 (2018).

Gaudillière, Jean-Paul, and Volker Hess, eds. *Ways of Regulating Drugs in the 19th and 20th Centuries*. Palgrave Macmillan, 2013.

Gilbert, Scott, Jan Sapp, and Alfred Tauber. "A Symbiotic View of Life: We Have Never Been Individuals." *Quarterly Review of Biology* 87, no. 4 (2012): 325–41.

Gillings, Michael. "Evolutionary Consequences of Antibiotic Use for the Resistome, Mobilome, and Microbial Pangenome." *Frontiers in Microbiology* 4, no. 4 (2013).

Gradmann, Christoph. "From Lighthouse to Hothouse: Hospital Hygiene, Antibiotics and the Evolution of Infectious Disease, 1950–1990." *History and Philosophy of Life Sciences* 40, no. 1 (2018).

Greene, Jeremy. *Prescribing by Numbers: Drugs and the Definition of Disease*. Johns Hopkins University Press, 2007.

Greene, Jeremy, Flurin Condrau, and Elizabeth Siegel Watkins, eds. *Therapeutic Revolutions: Pharmaceuticals and Social Change in the Twentieth Century*. University of Chicago Press.

Greenhough, Beth. "Where Species Meet and Mingle: Endemic Human–Virus Relations, Embodied Communication and More-Than-Human Agency at the Common Cold Unit 1946–90." *Cultural Geographies* 19, no. 3 (2012): 281–301.

Greenhough, Beth, Andrew Dwyer, Richard Grenyer, Timothy Hodgetts, Carmen McLeod, and Jamie Lorimer. "Unsettling Antibiosis: How Might Interdisciplinary Researchers Generate a Feeling for the Microbiome and to What Effect?" *Palgrave Communications* 4, no. 149 (2018).

Grote, Mathias, Lisa Onaga, Angela Creager, et al. "The Molecular Vista: Current Perspectives on Molecules and Life in the Twentieth Century." *History and Philosophy of the Life Sciences* 43, no. 1 (2021).

Hache, Émilie. *Ce à quoi nous tenons. Propositions pour une écologie pragmatique*. La Découverte, 2011.

Hacking, Ian. *Mad Travelers: Reflections on the Reality of Transient Mental Illness.* University of Virginia Press, 1998.

Haraway, Donna. "Anthropocene, Capitalocene, Plantationocene, Chthulucene: Making Kin." *Environmental Humanities* 6, no. 1 (2015): 159–65.

———. *The Companion Species Manifesto.* Chicago. University of Chicago Press, 2003.

———. *Modest_Witness@Second_Millenium.FemaleMan_Meets_OncoMouse: Feminism and Technoscience.* Routledge, 1997.

———. "Situated Knowledges: The Science Question in Feminism and the Privilege of Partial Perspective." *Feminist Studies* 14, no. 3 (1988): 575–99.

———. *Staying with the Trouble: Making Kin in the Chthulucene.* Duke University Press, 2016.

Hatfull, Graham. "Dark Matter of the Biosphere: The Amazing World of Bacteriophage Diversity." *Journal of Virology* 89, no. 16 (2015).

Häusler, Thomas. *Virus vs. Superbugs: A Solution to the Antibiotics Crisis?* Macmillan, 2006.

Helmreich, Stefan. *Alien Ocean: Anthropological Voyages in Microbial Seas.* University of California Press, 2009.

———. "Homo Microbis: The Human Microbiome, Figural, Literal, Political." *Threshold* 42 (2014): 52–59.

Hendy, Jessica, Matthäus Rest, Marl Aldenderfer, and Christina Warinner, eds. "Cultures of Fermentation." *Current Anthropology* 62, no. S24 (2021).

Hermitte, Marie-Angèle. *L'Emprise des droits intellectuels sur le monde vivant.* Quae, 2016.

Himmelweit, Freddy. "Combined Action of Penicillin and Bacteriophage on *Staphylococci.*" *Lancet* 246, no. 6361 (1945): 104–105.

Hinchliffe, Steve. "More Than One World, More Than One Health: Re-configuring Interspecies Health." *Social Science & Medicine* 129 (2015): 28–35.

Hinchliffe, Steve, Andrea Butcher, and Muhammad Meezanur Rahman. "The AMR Problem: Demanding Economies, Biological Margins, and Co-producing Alternative Strategies." *Palgrave Communications* 4, no. 142 (2018).

Holmes, Alison H., Luke S. P. Moore, Arnfinn Sundsfjord, et al. "Understanding the Mechanisms and Drivers of Antimicrobial Resistance." *Lancet* 387, no. 10014 (2016): 176–87.

Hooks, Katarzyna B., and Maureen O'Malley. "Dysbiosis and Its Discontents." *mBio* 8, no. 5 (2017): e01492–17.

Houdart, Sophie. *La Cour des miracles. Ethnologie d'un laboratoire japonais.* CNRS, 2008.

Hyman, Paul, and Stephen Abedon. "Bacteriophage Host Range and Bacterial Resistance." *Advances in Applied Microbiology* 70 (2010): 217–48.

Jault, Patrick, Thomas Leclerc, Serge Jennes, et al. "Efficacy and Tolerability of a Cocktail of Bacteriophages to Treat Burn Wounds Infected by *Pseudomonas aeruginosa* (PhagoBurn): A Randomised, Controlled, Double-Blind Phase 1/2 Trial." *Lancet Infectious Diseases* 19, no. 1 (2019): 35–45.

Kay, Lily. *The Molecular Vision of Life.* Oxford University Press, 1993.

Kameda, Koichi, and Charlotte Brives. "Reimagining Antimicrobial Innovation: Public Hospital-Based Therapy in Belgium and France (Forthcoming)

Kirchhelle, Claas. "The Forgotten Typers." *Notes and Records* 74, no. 4 (2019): 539–65.

——. *Pyrrhic Progress: The History of Antibiotics in Anglo-American Food Production.* Rutgers University Press, 2020.

Kirksey, Eben. "Queer Love, Gender Bending Bacteria, and Life After the Anthropocene." *Theory, Culture & Society* 36, no. 6 (2019): 197–219.

Kohler, Robert. *Lords of the Fly: Drosophila Genetics and the Experimental Life.* University of Chicago Press, 1994.

Koonin, Eugene, Pere Puigbò, and Yuri Wolf. "Comparison of Phylogenetic Trees and Search for a Central Trend in the 'Forest of Life.'" *Journal of Computational Biology* 18, no. 7 (2011): 917–24.

Kostyrka, Gladys. "La place des virus dans le monde vivant." PhD diss., Université Paris 1 Panthéon Sorbonne, 2018.

Krueger, Albert Paul, and Jane E. Scribner. "The Bacteriophage." *Journal of the American Medical Association* 116 (1941): 2160–69.

Kuchment, Anna. *The Forgotten Cure: The Past and Future of Phage Therapy.* Copernicus, 2012.

Lainé, Nicolas, and Serge Morand. "Linking Humans, Their Animals, and the Environment Again: A Decolonized and More-Than-Human Approach to 'One Health.'" *Parasite* 27, no. 55 (2020).

Landecker, Hannah. "Antibiotic Resistance and the Biology of History." *Body & Society* 22, no. 4 (2016): 19–52.

——. *Culturing Life: How Cells Became Technologies.* Harvard University Press, 2006.

———. "The Food of Our Food: Medicated Feed and the Industrialization of Metabolism." In *Eating Beside Ourselves: Thresholds of Foods and Bodies*, ed. Heather Paxson. Duke University Press, 2023.

Latour, Bruno. *Facing Gaia: Eight Lectures on the New Climate Regime*. Polity, 2017.

———. "Give Me a Laboratory and I Will Raise the World." In *Science Observed: Perspectives on the Social Study of Science*, ed. K. Knorr-Cetina and M. Mulkay. Sage, 1983.

———. *The Pasteurization of France*. Harvard University Press, 1988.

———. *We Have Never Been Modern*. Harvard University Press, 1993.

Latour, Bruno, and Steve Woolgar. *Laboratory Life: The Social Construction of Scientific Facts* (Sage, 1979).

Leitner, Lorenz, Aleksandre Ujmajuridze, Nina Chanishvili, et al. "Intravesical Bacteriophages for Treating Urinary Tract Infections in Patients Undergoing Transurethral Resection of the Prostate: A Randomised, Placebo-Controlled, Double-Blind Clinical Trial." *Lancet Infectious Diseases* 21, no. 3 (2020): 427–36.

Liboiron, Max. *Pollution Is Colonialism*. Duke University Press, 2021.

Lorimer, Jamie. *The Probiotic Planet: Using Life to Manage Life*. University of Minnesota Press, 2020.

Löwy, Ilana. "Martin Arrowsmith's Clinical Trial: Scientific Precision and Heroic Medicine." *Journal of the Royal Society of Medicine* 1 (2010): 461466.

Malm, Andreas. *L'Anthropocène contre l'histoire*. La Fabrique, 2018.

Margulis, Lynn, and Dorion Sagan. *Microcosmos: Four Billion Years of Microbial Evolution*. University of California Press, 1997.

Marks, Harry. "Confiance et méfiance envers le marché: les statistiques et la recherche clinique (1945–1960)." *Sciences sociales et santé* 18, no. 4 (2000): 9–27.

———. *The Progress of Experiment: Science and Therapeutic Reform in the United States 1900–1990*. Cambridge University Press, 1997.

Martin, Emily. "The Egg and the Sperm: How Science Has Constructed a Romance Based on Stereotypical Male–Female Roles." *Journal of Women in Culture and Society* 16, no. 3 (1991): 485–501.

———. *Flexible Bodies*. Beacon, 1995.

McCallin, Shawna, Shafiqul A. Sarker, Shamima Sultana, Frank Oechslin, and Harald Brüssow. "Metagenome Analysis of Russian and Georgian

Pyophage Cocktails and a Placebo-Controlled Safety Trial of Single Phage Versus Phage Cocktail in Healthy *Staphylococcus Aureus* Carriers." *Environmental Microbiology* 20 (2018): 3278–93.

McNeill, John Robert, and Peter Engelke. *The Great Acceleration: An Environmental History of the Anthropocene Since 1945.* Harvard University Press, 2014.

Merabishvili, Maya, Jean-Paul Pirnay, Gilbert Verbeken, et al. "Quality-Controlled Small-Scale Production of a Well-Defined Bacteriophage Cocktail for Use in Human Clinical Trials." *PLoS One* 4, no. 3 (2009): e4944.

Merchant, Carolyn. *The Death of Nature: Women, Ecology, and the Scientific Revolution.* HarperCollins, 1990.

Méthot Pierre-Olivier, and Samuel Alizon Samuel. "What Is a Pathogen? Toward a Process View of Host–Parasite Interactions." *Virulence* 5, no. 8 (2014): 775–85.

Milanovic, Fabien, ed. "Les ressources biologiques." *Revue d'anthropologie des connaissances* 2, no. 5 (2011).

Mintz, Sidney W. *Sweetness and Power: The Place of Sugar in Modern History.* Penguin, 1986.

Mol, Annemarie. *The Body Multiple: Ontology in Medical Practice.* Duke University Press, 2022.

——. *The Logic of Care: Health and the Problem of Patient Choice.* Routledge, 2008.

Monnin, Alexandre, Diego Landivar, and Emmanuel Bonnet. *Héritage et fermeture. Une écologie du démantèlement.* Divergences, 2021.

Morange, Michel. *Histoire de la biologie moléculaire.* La Découverte, 1994.

Morton, Harry, and Frank Engley. "Dysentery Bacteriophage: Review of the Literature on Its Prophylactic and Therapeutic Uses in Man and in Experimental Infections in Animals." *Journal of the American Medical Association* 127, no. 10 (1945): 584–91.

Moore, Jason W. *Capitalism in the Web of Life.* Verso, 2015.

——. "The Capitalocene, Part I: On the Nature and Origins of Our Ecological Crisis." *Journal of Peasant Studies* 44, no. 3 (2017): 594–630.

——. "The Capitalocene, Part II: Accumulation by Appropriation and the Centrality of Unpaid Work/Energy." *Journal of Peasant Studies* 45, no. 2 (2018): 237–79.

Moulin, Gérard, Philippe Cavalié, Isabelle Pellanne, et al. "A Comparison of Antimicrobial Usage in Human and Veterinary Medicine in France from 1999 to 2005." *Journal of Antimicrobial Chemotherapy* 62, no. 3 (2008): 617–25.

Murphy, Michelle. "Chemical Infrastructures of the Saint Clair River." In *Toxicants, Health and Regulation Since 1945*, ed. Soraya Boudia and Nathalie Jas. Pickering & Chatto, 2013.

Myelnikov, Dmitriy. "An Alternative Cure: The Adoption and Survival of Bacteriophage Therapy in the USSR, 1922–1955." *Journal of the History of Medicine and Allied Sciences* 73, no. 4 (2018): 385–411.

Nading, Alex. *Mosquito Trails: Ecology, Health, and the Politics of Entanglement.* University of California Press, 2014.

Nash, Linda. *Inescapable Ecologies: A History of Environment, Disease and Knowledge.* University of California Press, 2006.

Niewöhner, Jörg, and Margaret Lock. "Situating Local Biologies: Anthropological Perspectives on Environment/Human Entanglements." *BioSocieties* 13 (2018): 681–97.

O'Malley, Maureen. *Philosophy of Microbiology.* Cambridge University Press, 2014.

Orsi, Fabienne. "La constitution d'un nouveau droit de propriété intellectuelle sur le vivant aux États-Unis: origine et signification économique d'un dépassement de frontière." *Revue d'économie industrielle* 99 (2002): 65–86.

Patel, Raj, and Jason W. Moore. *A History of the World in Seven Cheap Things: A Guide to Capitalism, Nature and the Future of the Planet.* University of California Press, 2018.

Paxson, Heather. *The Life of Cheese: Crafting Food and Value in America.* University of California Press, 2012.

——. "Post-Pasteurian Cultures: The Microbiopolitics of Raw-Milk Cheese in the United States." *Cultural Anthropology* 23, no. 1 (2008): 15–47.

Paxson, Heather, and Stefan Helmreich. "The Perils and Promises of Microbial Abundance: Novel Natures and Model Ecosystems, from Artisanal Cheese to Alien Seas." *Social Studies of Science* 44, no. 2 (2014): 165–93.

Peduzzi, Peter, Martin Gruber, Michael Gruber, and Michael Schagerl. "The Virus's Tooth: Cyanophages Affect an African Flamingo Population in a Bottom-Up Cascade." *ISME Journal* 8, no. 6 (2014): 1346–51.

Pépin, Jacques. *The Origins of AIDS.* Cambridge University Press, 2011.

Petryna, Adriana. *When Experiments Travel: Clinical Trials and the Global Search for Human Subjects.* Princeton University Press, 2009.

Pignarre, Philippe. *Comment la dépression est devenue une épidémie*. La Découverte, 2012.

———. *Le Grand Secret de l'industrie pharmaceutique*. La Découverte, 2004.

Pirnay, Jean-Paul, Daniel De Vos, Gilbert Verbeken, et al. "The Phage Therapy Paradigm: Prêt-à-Porter or Sur-Mesure?" *Pharmaceutical Research* 28, no. 4 (2011): 934–37.

Pirnay, Jean-Paul, Gilbert Verbeken, Pieter-Jan Ceyssens, et al. "The Magistral Phage." *Viruses* 10, no. 2 (2018).

Podolsky, Scott H. *The Antibiotic Era: Reform, Resistance, and the Pursuit of a Rational Therapeutics*. Johns Hopkins University Press, 2015.

———. "Antibiotics and the Social History of the Controlled Clinical Trial, 1950–1970." *Journal of the History of Medicine and Allied Sciences* 65, no. 3 (2010): 327–67.

———. "The Evolving Response to Antibiotic Resistance (1945–2018)." *Palgrave Communications* 4, no. 124 (2018).

Podolsky, Scott H., Robert Bud, Christoph Gradmann, et al. "History Teaches Us That Confronting Antibiotic Resistance Requires Stronger Global Collective Action." *Journal of Law, Medicine and Ethics* 43, no. 2 (2015): 27–32.

Pourraz, Jessica. "Produire des médicaments en Afrique subsaharienne à l'heure de la santé globale. Le cas des antipaludiques au Ghana." *Politique africaine* 156, no. 4 (2019): 41–60.

Povinelli, Elizabeth A. *Geontologies: A Requiem to Late Liberalism*. Duke University Press, 2016.

Pradeu, Thomas, Gladys Kostyrka, and John Dupré. "Understanding Viruses: Philosophical Investigations." *Studies in History and Philosophy of Biological and Biomedical Sciences* 59 (2016): 57–63.

Proctor, Robert N., and Londa Schiebinger, eds. *Agnotology: The Making and Unmaking of Ignorance*. Stanford University Press, 2008.

Rader, Karen. *Making Mice: Standardizing Animals for American Biomedical Research, 1900–1955*. Princeton University Press, 2004.

Radin, Joanna, and Emma Kowal, eds. *Cryopolitics: Frozen Life in a Melting World*. MIT Press, 2017.

Rajan, Kaushik Sunder. *Biocapital*. Duke University Press, 2006.

———. *Pharmocracy: Value, Politics, and Knowledge in Global Medicine*. Duke University Press, 2017.

Rees, Tobias, Thomas Bosch, and Angela Douglas. "How the Microbiome Challenges Our Concept of Self." *PLoS Biology* 16, no. 2 (2018): e2005358.

Rohwer, Forest, Merry Youle, Heather Maughan, and Nao Hisakawa. *Life in Our Phage World: A Centennial Field Guide to the Earth's Most Diverse Inhabitants.* Wholon, 2014.

Rosebury, Theodor. *Microorganisms Indigenous to Man.* McGraw-Hill, 1962.

Sackett, David L., William M. C. Rosenberg, J. A. Muir Gray, R. Brian Haynes, and W. Scott Richardson. "Evidence-Based Medicine: What It Is and What It Isn't." *British Medical Journal* 312 (1996): 71–72.

Sagan, Lynn. "On the Origin of Mitosing Cells." *Journal of Theoretical Biology* 14, no. 3 (1967): 225–74.

Sanabria, Emilia. "Circulating Ignorance: Complexity and Agnogenesis in the Obesity 'Epidemics.'" *Cultural Anthropology* 31, no. 1 (2016): 131–58.

——. *Plastic Bodies: Sex Hormones and Menstrual Suppression in Brazil.* Duke University Press, 2016.

Sangodeyi, Funke Iyabo. "The Making of the Microbial Body, 1900s–2012." PhD diss., Harvard University, 2014.

Schindler, Thomas E. *A Hidden Legacy: The Life and Work of Esther Zimmer Lederberg.* Oxford University Press, 2021.

Selosse, Marc-André. *Jamais seul. Ces microbes qui construisent les plantes, les animaux et les civilisations.* Actes Sud, 2017.

Servitje, Lorenzo. *Medicine Is War: The Martial Metaphor in Victorian Literature and Culture.* State University of New York Press, 2021.

Silbergeld, Ellen. *Chickenizing Farms and Food.* Johns Hopkins University Press, 2016.

Singer Andrew, Claas Kirchhelle, and Adam Roberts. "Reinventing the Antimicrobial Pipeline in Response to the Global Crisis of Antimicrobial-Resistant Infections." *F1000 Research* 8 (2019): 238.

Spagnolo, Fabrizio, Monica Trujillo, and John Dennehy. "Why Do Antibiotics Exist?" *mBio* 12, no. 6 (2021): e01966–21.

Star, Susan Leigh. "The Ethnography of Infrastructure." *American Behavioral Scientist* 43, no. 3 (1999): 377–91.

Steffen, Will, Paul J. Crutzen, and John R. McNeill. "The Anthropocene: Are Humans Now Overwhelming the Great Forces of Nature?" *Ambio* 36, no. 8 (2007): 614–21.

Stekel, Dov. "First Report of Antimicrobial Resistance Pre-dates Penicillin." *Nature* 562, no. 7726 (2018): 192.

Stengers, Isabelle. *In Catastrophic Times: Resisting the Coming Barbarism.* Open Humanities, 2015.

——. *The Invention of Modern Science.* University of Minnesota Press, 2000.

——. *Résister au désastre.* Wildproject, 2019.

——. *La Vierge et le Neutrino. Les scientifiques dans la tourmente.* Les Empêcheurs de penser en rond, 2006.

Strathdee, Steffanie, and Thomas Patterson. *The Perfect Predator: A Scientist's Race to Save Her Husband from a Deadly Superbug.* Hachette, 2019.

Strathern, Marylin. *Partial Connections.* AltaMira, 1991.

Summers, William C. "Cholera and Plague in India: The Bacteriophage Inquiry of 1927–1936." *Journal of the History of Medicine and Allied Sciences* 48 (1993): 275–301.

——. "On the Origins of the Science in *Arrowsmith*: Paul de Kruif, Félix d'Hérelle, and Phage." *Journal of the History and Medicine and Allied Sciences* 46 (1991): 315–32.

——. "The Strange History of Phage Therapy." *Bacteriophage* 2, no. 2 (2012): 130–33.

Suttle, Curtis. "Marine Viruses—Major Players in the Global Ecosystem." *Nature Reviews* 5, no. 10 (2007): 801–12.

Tang, Karen L., Niamh P. Caffrey, Diego B. Nóbrega, et al. "Restricting the Use of Antibiotics in Food-Producing Animals and Its Associations with Antibiotic Resistance in Food-Producing Animals and Human Beings: A Systematic Review and Meta-analysis." *Lancet Planetary Health* 1, no. 8 (2017): e316–27.

Thoreau, François. "L'embarquement par son objet, trois politiques de l'enquête sur les clôtures virtuelles (*virtual fences*)." *Revue d'anthropologie des connaissances* 13, no. 2 (2019): 399–423.

Timmermans, Stefan, and Marc Berg. *The Gold Standard: The Challenge of Evidence-Based Medicine and Standardization in Health Care.* Temple University Press, 2003.

Tsing, Anna. *The Mushroom at the End of the World: On the Possibility of Life in Capitalist Ruins.* Princeton University Press, 2015.

——. *Proliférations.* Wildproject, 2022.

Van Dooren, Thom. *Flight Ways: Life and Loss at the Edge of Extinction.* Columbia University Press, 2014.

Van Dooren, Thom, Eben Kirksey, and Ursula Münster. "Multispecies Studies: Cultivating Arts of Attentiveness." Special issue, *Environmental Humanities* 8, no. 1 (2016): 1–23.

Verbeken, Gilbert, Daniel De Vos, Mario Vaneechoutte, Maya Merabishvili, Martin Zizi, and Jean-Paul Pirnay. "European Regulatory Conundrum of Phage Therapy." *Future Microbiology* 2, no. 5 (2007): 485–91.

Villarroel, Julia, Larsen Mette Voldby, Mogens Kilstrup, and Morten Nielsen. "Metagenomic Analysis of Therapeutic PYO Phage Cocktails from 1997 to 2014." *Viruses* 9, no. 11 (2017): 328.

Wald, Priscilla. *Contagious: Cultures, Carriers, and the Outbreak Narrative.* Duke University Press, 2008.

Waldby, Catherine, and Robert Mitchell. *Tissue Economies: Blood, Organs, and Cell Lines in Late Capitalism.* Duke University Press, 2006.

Whyte, Susan Reynolds, Sjaak van der Geest, and Anita Hardon, eds. *Social Lives of Medicine.* Cambridge University Press, 2002.

Wilhelm, Steven, and Curtis Suttle. "Viruses and Nutrient Cycles in the Sea." *Bioscience* 49, no. 10 (1999): 781–88.

Woods, Abigail. "Decentring Antibiotics: UK Responses to the Diseases of Intensive Pig Production (ca. 1925–1965)." *Palgrave Communications* 5, no. 41 (2019).

World Health Organization. *Global Action Plan on Antimicrobial Resistance.* World Health Organization, 2014.

Wright, Gerard D., and Arlene D. Sutherland. "New Strategies for Combating Multidrug-Resistant Bacteria." *Trends in Molecular Medicine* 13, no 6 (2007): 260–67.

Yong, Ed. *I Contain Multitudes: The Microbes Within Us and a Grander View of Life.* HarperCollins, 2016.

Youle, Merry. *Thinking Like a Phage: The Genius of the Viruses That Infect Bacteria and Archaea.* Wholon, 2017.

Zimmer, Alexis. *Brouillards toxiques. Vallée de la Meuse, 1930, contre-enquête.* Zones sensibles, 2016.

——. "Collecter, conserver, cultiver des microbiotes intestinaux. Une biologie du sauvetage" *Écologie & politique* 58 (2019): 135–50.

———. "The Disappearing Microbiota: the Coloniality of a Narrative and Anti-Colonial Proposals," *Environmental Humanities* 17, no. 2 (2025): 351–70.

Zitouni, Benedikte. "Héritières de la Révolution scientifique: d'autres figures et manières de faire science." *Nouvelles questions féministes* 40, no. 2 (2021): 35–51.

Zschach, Henrike, Katrine Joensen, Barbara Lindhard, et al. "What Can We Learn from a Metagenomic Analysis of a Georgian Bacteriophage Cocktail?" *Viruses* 7, no. 12 (2015): 6570–89.

INDEX

GPSR Authorized Representative: Easy Access System Europe, Mustamäe tee
50, 10621 Tallinn, Estonia, gpsr.requests@easproject.com

www.ingramcontent.com/pod-product-compliance
Lightning Source LLC
Chambersburg PA
CBHW021856020426
42334CB00013B/351